SELECTED LITERARY CRITICISM OF LOUIS MACNEICE

Selected Literary Criticism of Louis MacNeice

Edited by
ALAN HEUSER

CLARENDON PRESS · OXFORD
1987

PR
6025
A316
A6
1987

Oxford University Press, Walton Street, Oxford OX2 6DP
Oxford New York Toronto
Delhi Bombay Calcutta Madras Karachi
Petaling Jaya Singapore Hong Kong Tokyo
Nairobi Dar es Salaam Cape Town
Melbourne Auckland
and associated companies in
Beirut Berlin Ibadan Nicosia

Oxford is a trade mark of Oxford University Press

Published in the United States by
Oxford University Press, New York

© *Copyright of text by The Estate of Louis MacNeice 1987*
© *Copyright of introduction, notes, and bibliography*
by Alan Heuser 1987

All rights reserved. No part of this publication may be reproduced, stored in a retrieval system, or transmitted, in any form or by any means, electronic, mechanical, photocopying, recording, or otherwise, without the prior permission of Oxford University Press

British Library Cataloguing in Publication Data
MacNeice, Louis
Selected literary criticism of Louis MacNeice.
1. English literature—History and criticism
I. Title II. Heuser, Alan
820.9 PR83
ISBN 0-19-818573-1

Library of Congress Cataloging in Publication Data
MacNeice, Louis, 1907-1963.
Selected literary criticism of Louis MacNeice.
Bibliography: p.
Includes index.
1. Books—Reviews. 2. Literature—History and criticism. 3. Poetry. I. Heuser, Alan.
PR6025.A316A6 1987 809 86-16409
ISBN 0-19-818573-1

Phototypeset by Dobbie Typesetting Service, Plymouth, Devon
Printed in Great Britain
at the University Printing House, Oxford
by David Stanford
Printer to the University

To Margaret and Dan

Acknowledgements

MANY people, my university, and my national research council have facilitated the work of this book and a planned second volume. First, thanks are due to the Humanities Research Committee of McGill University, whose travel grant in 1981 enabled me to consult the large Louis MacNeice Collection of Manuscripts at the Humanities Research Center, Austin, Texas. Then, for research grants in 1982 and 1985 taking me to England to consult the various collections of Louis MacNeice Papers at the Bodleian Library, Oxford, with further research in Reading, London, and Manchester—as well as for a leave fellowship 1984–5—I have to thank the Social Sciences and Humanities Research Council of Canada, Ottawa. Moreover, gratitude is due to McGill University for granting me sabbatical leave in 1984–5 to complete this book, and especially to David Williams, Chairman of the English Department, for facilitating the enterprise.

For permission to publish Louis MacNeice's prose, grateful acknowledgement is made to the literary executor Dan Davin, to Mrs Hedli MacNeice, Daniel MacNeice, and the MacNeice estate, and to the literary agents David Higham Associates.

For primary materials, the following were a great help: bibliographies of MacNeice by William T. McKinnon, *Bulletin of Bibliography* (1970), and by C. M. Armitage and Neil Clark (1973), the Auden bibliography by B. C. Bloomfield and Edward Mendelson (2nd edn, 1972), Barbara Coulton's *Louis MacNeice in the BBC* (1980), a letter from Professor Mendelson. For photocopies of some of MacNeice's manuscripts and typescripts, I wish to thank: Mrs Hedli MacNeice; Mrs Ellen S. Dunlap of the HRC, Austin; Mrs Edna Longley of the Queen's University of Belfast; Mrs Lola L. Szladits of the Berg Collection, New York Public Library; Robert A. Tibbetts of Ohio State University. For photocopies of MacNeice's published prose, thanks are due to: Miss Robyn Marsack; Mrs Jacqueline Kavanagh and Miss Gwyniver Jones of the BBC Written Archives Centre, Reading; A. J. Flavell of the Bodleian Library; E. G. H. Kempson of Marlborough College; John Griffin of the British Library; John R. Whitehead of Moffat, Dumfriesshire; Sir William Deedes of the *Daily Telegraph*; Kevin Andrews of Athens. Terence Kilmartin kindly sent a list of MacNeice's book reviews in the *Observer* 1949–62. Charles Seaton graciously opened for me the records of contributors to the *Spectator* up to 1938, only after which authors were fully indexed. Philip Howard of the *Times* and Jeremy Treglown of *TLS* were very hospitable, while the archivists Gordon Phillips and Mrs Anne Piggott were helpful. Rowley Newton assisted in the initial research in *Radio Times*, Martin Lubowski in *Radio Times*, *London Calling*,and *Time & Tide*.

I am grateful to the following persons for interviews: Walter Allen, Seamus Heaney, Sir Stephen Spender, Anthony Thwaite, John Whitehead, Michael Yates. Willing help was offered by the librarians of Broadcasting House, BBC Written Archives Centre, the British Institute of Recorded Sound, the British Library, the British Newspaper Library (Collindale), by Peter McNiven of The John Rylands University Library of Manchester, and especially by reference and interlibrary-loans librarians of McLennan Library, McGill University, as well as by my colleagues Abbott Conway, Archie Malloch, Albert Schachter.

For assistance in tracing some references and quotations I wish to thank Walter Allen, Jonathan Barker, D. S. Porter and A. J. Flavell of the Bodleian, the Rt. Hon. John Freeman, Jasper Griffin, John Jordan of the BBC Written Archives Centre, Professor R. J. Finneran, Professor Edward Mendelson, my colleagues Chris Heppner, the Rev. George Johnston, Hereward ('Wake') Senior, and Alain Tichoux, although I accept responsibility for any errors. In addition to all of the above, I am grateful for correspondence to Dr. Benedict S. Benedikz, Sir Isaiah Berlin, Gillian Boyd, Julie Carpenter, John Fuller, Roy Fuller, James H. Gray, the late Geoffrey Grigson, John Hilton, John Lehmann, Kenneth Lohf, Ronald Macfarlane, John S. North, Anthony B. Rota, Michael J. Sidnell, the late Reggie D. Smith, Terence Tiller, John Trew, the late Rex Warner, Mrs Eleanor Clark Warren. For assistance in proof-reading and the preparation of an index, I have to thank Susan Keys.

For encouragement over many years I am grateful to Joyce Hemlow. At all times Dan Davin and Jon Stallworthy were available with advice and practical help, while for encouragement, support, and understanding I owe the largest debt to my family, Margaret, Liesel, and John.

McGill University A. H.
July 1986

Contents

Introduction

LOUIS MACNEICE, the Anglo-Irish poet and liberal humanist, the Oxonian scholar and BBC playwright, a gregariously lonely man, reticent, but with the Irish gift of the gab when he wanted to use it, was a challengingly prolific writer who produced eighteen volumes of poetry, two notable verse translations that appeared in book form, a great many plays and features, ten books of prose, many articles and contributions to books, and a vast amount of journalism as well as unpublished material. His extensive and published occasional prose-writing from 1924 to 1963, printed as 'A Bibliography of Short Prose' at the back of this book, includes well over 300 items, of which more than half are 'new' and have not been listed before; they can be found in such periodicals as the *Spectator* (before full indexing of authors began in 1939), the *Listener* (unsigned reviews), the *Morning Post*, the *Observer*, *Time & Tide*, *Common Sense*, *Radio Times*, *London Calling* (these are not the only new periodical listings but the outstanding ones). Searches in the records of a few periodicals could not be made, though no unsigned items turned up in *TLS* and no items in the *Manchester Guardian*, the Belfast *News Letter*, the *Belfast Telegraph*, nor the *Anglo-Hellenic Review* (Athens).

The Bibliography includes *all* the materials of short prose traced so far and is not limited to literary criticism. Another selection of more varied prose is planned to be published in two years—reviews of and comments on philosophy, history, travel, Celtic matters (especially Ireland), 'London Letters' written during the Second World War, rugby, some autobiography, and several lively parodies.

The purpose of this book and of the subsequent volume is to publish two selections of the scattered and hitherto-uncollected short prose of MacNeice. Rules for exclusion in the present book have had to be rigorous: no juvenilia (that is, nothing before 1930), no strictly manuscript materials, no reviews of fiction, film, television, no journalism subsequently absorbed into his books, no introductions to nor selections from his own books, no broadcasts nor radio writing (nothing from *Radio Times* or

London Calling), and as little overlapping as possible—with the result that the avid reader may well find the Bibliography of more importance than the selection offered. Even with such stringent exclusions, there remains too much occasional literary criticism for a single annotated volume limited to 100,000 words by the publisher, so principles for inclusion have had to be reasonable and discriminating: articles of special interest, contributions to books not his own, a generous selection of classical reviews (though no philosophy), emphasis upon articles and reviews on poetry, especially on the four important poets he knew and admired—his seniors W. B. Yeats and T. S. Eliot, his contemporaries W. H. Auden and Dylan Thomas. Moreover, a certain balance between early and late work had to be kept in mind, so that in spite of the large bulk of early material and the great length of a few of these articles, the later work would not be scanted.

MacNeice never collected his critical articles and reviews probably because he considered them occasional and did not take time to revise; the ongoing process of writing poetry and plays was more important. He could never settle for any system, for those Procrustean tendencies and categories which some find necessary to the practice of professional literary criticism: he was constantly tilting against such follies. His three critical books were remarkable examples of breaking new ground, explorations from a poet's personal experience: *Modern Poetry* (1938) defended the new poetry from a case-book of his own poetic development; *The Poetry of W. B. Yeats* (1941), the first book of its kind, explained Yeats's poetic integrity from firsthand knowledge of æstheticism and of Ireland; *Varieties of Parable* (1965) explored, honestly and without neat system, the puzzling subject of double-level writing—from Spenser to Pinter—out of his experience of writing parable plays and a persistent dream life. All these were subjects close enough and new enough to engage his creative attention. His occasional criticism, fresh in its time and most of it still fresh, was written as notes on the way, *obiter dicta* thrown out in the forward movement and practice of his craft. If he had lived to collect the materials of such a book as this, he would of course have seriously revised them, eliminating some things, including a few repetitions unavoidable here, and adding others.

MacNeice's world may have been one of non-transcendence, of prose,[1] a common-sense earthy reality, yet it was transformed by psychic types and scenes out of dream, fantasy, myth—the imagination alive within him from childhood, from boyhood readings of saga, epic, romance, and the picaresque. Anchored in the facts of the waking life but prey to dream and nightmare, he recognized the uncanny dream logic of story in such works as Malory's *Morte Darthur*, the *Njals Saga*, the *Odyssey*, *The Faerie Queene*, *The Golden Ass*, the *Satyricon*.[2] Mythic pattern and dream rhythm penetrated the prison of an honest prose world with hovering psychic truths—either as paradoxes or anomalies or as genuine double vision—which he was keen to point out and defend, from an early article on Malory, through his Introduction to *The Golden Ass*, to a late review of Spenser. The article on Malory (April 1936) indicates how 'the dialectic of the dream' becomes 'concrete', how Malory's characters are 'real', how Sir Gawaine 'Of all the more important knights . . . is the furthest from being a paragon', how therefore he is a fascinating, flawed character.[3] In Apuleius' *Golden Ass* (1946) MacNeice as connoisseur of paradox was refreshed by a writer who combined 'elegance and earthiness, euphuism and realism, sophistication and love of folk-lore, Rabelaisian humour and lyrical daintiness, Platonism and belief in witchcraft, mysticism and salty irony'. His ironic mind loved to play with dialectic and paradox. In Spenser (January 1963) he found vindication for both his dream life and its inner psychic logic, being flexible enough to entertain Graham Hough's 'diagrammatic scheme' for 'the whole of literature' as a convenient system—something he had

[1] 'MacNeice employs great poetic gifts, and projects a genuinely poetic personality, to surround ideas and situations that are essentially those of prose. Although in many of his qualities he is superior to the world, he never ceases to belong to it': Stephen Spender, 'Songs of an Unsung Classicist' (review of New York reprint of MacNeice's *Collected Poems 1925–1948*), *Saturday Review* 66 (7 Sept. 1963), 25. See also Sir Stephen's 'The Brilliant Mr MacNeice', *New Republic* 156: 4 (28 Jan. 1967), 32–4.

[2] Many of these are mentioned in MacNeice's autobiography *The Strings are False* (1965); all but Malory were made into plays, readings, or adaptations for the BBC in MacNeice's productions 1944–60.

[3] See 'Landscapes of Childhood and Youth', *The Strings are False*, 221, for MacNeice's impersonation of Sir Gawaine at school as a model he could identify with. For further personal information, see the forthcoming biography of MacNeice by Jon Stallworthy.

rarely done—because it sparked his creative spirit in writing the Clark Lectures which were to make up his last book *Varieties of Parable*; but of course he did not systematize his own findings more than chronologically.

We are fortunate that MacNeice, a double-first classics scholar, reviewed two actual productions of Sophocles in Greek: the *Oedipus Tyrannus* acted by the boys of Bradfield with music by Abdy Williams (June 1937) and the *Antigone* acted by the undergraduates of Cambridge under the direction of J. T. Sheppard (March 1939). That was some time after the Group Theatre production of his own English verse *Agamemnon* (November 1936). Knowing this last fact helps one understand how he must have envied 'the grand thing' at Bradfield ('that the chorus can be put where they belong—in the *orchestra*': on the first night, his chorus had been only too conspicuous onstage, in dinner jackets) and the impressive set at Cambridge: 'a palace of Mycenean blocks'.

In matters of poetic translation, following his mentor Professor E. R. Dodds, MacNeice insisted on the integrity of the verse-line, maintaining a translator should give a line-by-line translation from the word order of the original wherever possible. In an early review (May 1935), while Gilbert Murray was still Regius Professor of Greek at Oxford, MacNeice attacked Murray for violating this principle in his Aeschylus' *Seven Against Thebes*, as well as for lacking 'technical virility'![4] A decade later (May 1945), while commending R. C. Trevelyan for observing the line-by-line principle in his Virgil, MacNeice's discriminating ear detected faults in Trevelyan's metrics or 'mastery of movement'. Still later (October 1962) he criticized Robert Fitzgerald's *Odyssey* for losing the tautness required in a long poem by not observing line-by-line translation[5] and for 'using a five-stress basically iambic line, though he often uses it felicitously': the resounding Homeric hexameter with its dactyls had been violated. As a practitioner of verse translation in his *Agamemnon* and *Faust*, as well as in translations of Horace and others, he vindicated such detailed

[4] Of course, MacNeice, under Dodds, was translating Aeschylus at this time. E. R. Dodds became Regius Professor of Greek at Oxford in 1936, succeeding Murray.

[5] Over a decade before this (i.e. before 1962), MacNeice had written laboriously with Ernst Stahl their English *Faust* (1951, 1965).

criticism by his own example of astonishing metrical virtuosity maintaining as much of the poetic integrity of the original as he could, without archaism.

When it came to twentieth-century poetry for a writer of such exacting standards, MacNeice, though no critic's fool, was limited by unwillingness to read much modern American verse: the American Imagists had discouraged him with their miniature poems of arbitrary and often slack free verse. His criticism of modern American poets excluding Pound and Eliot, Frost and Cummings, may be scanty, but his critiques of these four poets were as usual full of insight from firsthand reading. Within the range of English, Scottish, and Irish poets his discriminations were keen. Although he knew many of the American poets, he could not catch the point and tune of Wallace Stevens.

Two early long articles—'Poetry To-day' (September 1935) and 'Subject in Modern Poetry' (December 1936)—are not mere foretastes of *Modern Poetry* (1938) but well worth reading for their own sake, full of cogent distinctions and explanations of the new poetry for the reading public. His combination of critical insights and home truths from his own poetic practice was remarkable enough for an editor, E. B. Osborn, to praise the 1935 article as 'the most brilliant piece of writing' in Geoffrey Grigson's *The Arts To-Day*,[6] a book also featuring Auden. The next year MacNeice was invited to give an English Association lecture—the 1936 article—and then commissioned by the Oxford University Press to write *Modern Poetry*.

Three articles of 1948–9, long neglected but perhaps even better criticism than these of 1935–6, stand in significant relation to MacNeice's later creative work. 'An Alphabet of Literary Prejudices' (March 1948) tilts at various literary targets in a manner at once provocative and reasonable. 'Experiences with Images' (1949) testifying to the return in poetry, inspired by T. S. Eliot, 'to Shakespeare's catholic receptivity', shows some of his own early poetic attempts in an 'exercise in humility'.[7] The third of these articles, 'Poetry, the Public, and the Critic' (October

[6] The *Morning Post*, 10 Sept. 1935, 14.

[7] MacNeice had already defended Eliot's technical expertise by citing similar supposed 'bad' examples to be found in Shakespeare, in MacNeice's article 'The Traditional Aspect of Modern English Poetry' (Dec. 1946).

1949), is a fine critical piece of sound sense written in plain prose: the average reader 'cannot be *poetically* educated until we have responsible critics'. He had just published his *Collected Poems 1925–1928* (September 1949) and knew from past experience that many reviewers were not educated in the inner workings of poetry.

About his seniors Yeats and Eliot MacNeice had remarkably little anxiety of influence; he admired and did otherwise. Yeats, the great Irish poet ever renewing his poetry, was a good prospect to review, even posthumously. MacNeice found Synge's 'astringent joy and hardness' decisive in the later Yeats, and the latter had become 'the best example of how a poet ought to develop if he goes on writing till he is old' ('Poetry To-day', September 1935). Yeats's death and *Last Poems and Plays* prompted a review (June 1940), then another Oxford book, *The Poetry of W. B. Yeats* (1941). Reviewing Yeats's *Collected Poems* (August 1950) MacNeice hoped for a complete variorum edition; when that appeared he further hoped (December 1958) for a study of the manuscripts: when Jon Stallworthy's study of Yeats's poetic work-sheets appeared subsequently, he reviewed it favourably (March 1963), especially approving Yeats's habit of revision by cutting. So for the scholar-poet-critic scholarship and criticism were forward moving as well as modern poetry.

Eliot's verse and prose of 1917–33 were paramount and instructive to MacNeice: he tended to take a retrospective view, for 'Prufrock' and *The Waste Land* had left deep imprints on him in his formative years which Eliot's later verse could not supersede; and Eliot remained the great Cham whose critical dicta MacNeice found useful in his own criticism.[8] Through his fine ear for rhythm and tone MacNeice appreciated Eliot's technical experiments and recognized that he had widened the range of subject matter in English poetry and increased its precision ('Subject in Modern Poetry', December 1936). Whereas MacNeice described the typical Yeats play as undramatic with only lyric intensity, he applauded Eliot's advances with more supple verse in the more dramatic form (May 1939, May 1950).

[8] MacNeice most frequently cited and quoted from 'Tradition and the Individual Talent' (1919), 'The Metaphysical Poets' (1921), and *The Use of Poetry and the Use of Criticism* (1933), among Eliot's critical works.

MacNeice's contemporaries Wystan Auden and Dylan Thomas were poet-friends close to him at different periods of his life: Auden 1930–40, Thomas 1942–53. Early on he saluted Auden's forthright manner of the 'poem-telegram' (March 1931), his startling development, his 'mix-up of politics and psychology' (December 1936); and after the publication of their joint *Letters from Iceland* (August 1937) he wrote a letter in *New Verse* (November 1937) warning him against crude ballads, 'writing down to the crowd', and too much realism. Reporting on Auden and Isherwood's play *On the Frontier* (November 1938), he recorded a distinct embarrassment: 'The mystical love scenes of Eric and Anna made one long for a sack to put one's head in'. His last review of Auden's poetry (April 1940)—written while teaching in the USA—recognized in Auden an 'astonishing versatility' and continuing integrity which survived through Auden's adoption of 'semi-mystical' religious orthodoxy along with psychological dialectic still 'not defeatist'.[9]

All but one of MacNeice's articles on Dylan Thomas—the review of the *Collected Poems* (April 1953)—were written after that poet's death. He always distinguished between Thomas's poetic power and the bad poetry of his imitators. After Thomas's rather sad end he saw him even more clearly as a bard 'with the three great bardic virtues of faith, joy, and craftsmanship' (January 1954); was amused that the man called 'The Poet of the Pubs' could never bring himself 'to use the word "pub" in a poem' (April 1954), so controlled and limited was he by his own poetic diction; countered the spiteful gossip about Thomas's 'Bohemian' qualities by describing him as 'a comic genius—and the cause of comedy in others' (*Ingot*, December 1954) with a charming balance of wit and humour evident in his prose (*New York Times Book Review*, December 1954).

Among later articles and reviews there are lively comments on the Movement poets and 'Movementese' (April 1957); on 'the growing tribes of humourless scholars' studying the three Irishmen Shaw, Yeats, and Joyce (November 1957); on Enid

[9] MacNeice often regarded T. S. Eliot, at least in his earlier poetry, as 'defeatist', and the 'reaction' of his own generation of poets as partly against Eliot. After his return to England from the USA, MacNeice was quick to defend Auden and other English expatriate writers for their American residence during the war ('Traveller's Return', *Horizon*, Feb. 1941).

Starkie's king-making forays for the Poetry Chair at Oxford (February 1961); on the advantage of reading merely informative books—a practice not 'tainted by any hint of the snob game' of æsthetic judgement (August 1961); and three important short articles on his last three volumes of poetry (May 1957, February 1961, September 1963).

Such are only some of the highlights in the selection.[10] A distinct drawback of my rules for exclusion is the omission of MacNeice's comments on his radio plays and features in the *Radio Times* : there is just not space enough here; perhaps a slim volume of these might be published in the future.

It should become clear from the material edited in this book, and from his works as a whole, that MacNeice was a very perceptive critic.[11] His quick, allusive intelligence penetrated with delicate and humble tact to the essentials of his reading, especially in matters of technique, undistracted by snob values, labels, the pigeon-holing of more conventional literary criticism. His critical prose was as vital and readable as it was delicate and tactful, though he could be witty or dry when ironic playfulness was called for. Honesty, 'dissatisfaction with accepted formulas' (October 1934), 'the assertion of human values' ('Eclogue from Iceland'), absence of pose, pragmatic probing to note both unities and inconsistencies, a wide humanity and the appreciation of 'appetitive decorum' ('Memoranda to Horace' iv) were constants in his development as critic, though he gave up some of the brilliant flourishes of his early manner for plainer, deeper lights in his later style, always able to take delight in the 'incorrigibly plural' world ('Snow') and the paradoxes of the human condition. If

[10] What has been left out includes such miscellaneous figures and works as Cocteau (Feb. 1937); Lorca, and Waley's Chinese Poems (both Oct. 1937); Hans Christian Andersen (Jan. 1938); Housman (April 1940); Keats (Jan. 1941); O'Casey's memoirs (Nov. 1945, Feb. 1949, July 1952, May 1963); the New Testament of Ronald Knox (May 1946); Byron (April 1949); Betjeman (May 1949); Augustus John (April 1952); Edward Lear (April 1953); E. M. W. Tillyard (June 1954); Burns and Clare (Aug. 1956); Wilfred Owen (Oct. 1960); Frank O'Connor (June 1961); William Empsom (1963). Among MacNeice's editors were Geoffrey Grigson of *New Verse*, T. S. Eliot of the *Criterion*, J. R. Ackerley of the *Listener*, Wilson Harris of the *Spectator*, Cyril Connolly of *Horizon*, John Lehmann of *New Writing* and the *London Magazine*, Jim Rose and Terence Kilmartin of the *Observer*. It is not possible to name all.

[11] For a sympathetic account, see John Wain, 'MacNeice as Critic', *Encounter* 27: 5 (Nov. 1966), 49–55: 'Louis MacNeice, like most poets, was a good critic, and his criticism seems to me to have been unjustly neglected.'

for MacNeice the poet is a concentration of the ordinary man (Everyman) who rediscovers commonplaces ('The Truisms') as a good maker or craftsman, the critic who is also a practitioner of the craft presents creative principles which arise from his own quest of life and form.

A Note on the Text

Because of the many varying house styles used over three decades (1930–63) and on both sides of the Atlantic by the journalistic and press editors of MacNeice's prose, it has been thought best to standardize (and, for a few American spellings, Anglicize) MacNeice's text according to current style: thus verbs and compounds ending in -ise have been altered to end in -ize, Mr. to Mr, mediæval to medieval, and so on. The one exception is to-day, which was printed as such in the vast majority of instances. Also, short poems are given in roman type and inverted commas, instead of in italics, which are reserved for book-titles. In footnotes, whenever two or more dates are cited together, dates of writing are given outside parentheses immediately after titles, whereas parentheses after titles of works and their dates of writing are reserved for dates of publication. The footnotes are intended for the common reader as well as the scholar.

Poems, by W. H. Auden[1]

Poems, by W. H. AUDEN. (Faber & Faber, 2*s*. 6*d*.)

God (or Nature) has a diffuse style which poets have often been busied correcting. Especially modern poets. Mr Auden's attempt is to put the soul across in telegrams. But whereas in the everyday telegram the words tend to be, like Morse, mere counters, in the poem-telegram the words stand rather on their own than *for* a meaning behind them. Many would consider much in these poems to be irrelevant, arbitrary, indifferent.

But in all poetry one feels rather than knows the irrelevance and relevance. Only in new poetry one tends to start knowing in advance of feeling, because the poems come to one rather as curios and samples than as something valuable. And critics, too, are *qua* critics bound to consider what is curious or typical. Which is why I would rather be a propagandist than a critic.

Mr Auden, then, uses an up-to-date technique to express an up-to-date mood (not that either 'use' or 'technique' or 'express' or 'mood' has, in this sentence, any legitimate sense). Generalizations of this sort are no good except as rough hints. And special instances are little better. To detect an 'instance' of the influence of Robert Graves is little, perhaps, better than an instance of the conditional sentence. A good poet is no more a conglomeration of the typical and the derivative (and the peculiar) than of the grammatical. As for bad poets—there are also people who write solely to write grammar.

Still, though both are conditions and not causes, contemporary literary influence is a more changing condition than grammar. Which is why it may slightly elucidate these poems (while having nothing to do with their value) if one suggests that Mr Auden is well-read in the 'typical' 'advanced' reading of to-day, that having been helped to see things newly by modern psychology he is helped to present them in a new and strenuous presentation

[1] *Oxford Outlook* 11: 52 (March 1931), 59–61. Auden's *Poems* were published by Faber 18 Sept. 1930.

by, among others, Eliot, Robert Graves, the later Yeats, G. M. Hopkins and Wilfren Owen.[2]

We must beware of facile analysis. When Mr Auden says:

> Gannets blown over northward, going home,
> Surprised the secrecy beneath the skin,[3]

we must not narrow and so destroy his 'gannets'—either by narrowing them to meaning (as the scientist) or to sound-value (as, sometimes, Edith Sitwell).[4] We can merely say that gannet is the right sound (and, incidentally, an Eliot sound) and that the sound adds to the meaning and by adding makes a new meaning (= makes poetry = creates). And it is unjust when Mr Auden uses a word like 'sessile' to say he uses it because it is or sounds 'scientific'; more likely he uses it just because it sounds. Mr Auden has a remarkable ear which he runs in harness with his mind.

A critic is forced to insult poetry by discussing its 'elements' separately. So now for his 'mood'. Mr Auden's mood is that notorious disillusionment of our time which yet goes forward to realize itself—along roads which its own bombs have riddled (e.g. Lawrence recognizing sexual breakdown finds Heaven in sex; Eliot seeing the vanity of polymathy compiles a solid from vanities; Cummings by and in disintegrating, aims at new integrities).[5] So Mr Auden to pin the flux does not mangle it into something else, but tries to pin it as it is and yet to pin it, e.g.:

> Is first baby, warm in mother,
> Before born and is still mother,

[2] All these were poets much read in the 1920s and esp. in 1928–31: T. S. Eliot (1888–1965), from *Prufrock* (1917) to *Ash-Wednesday* (1930); Robert R. Graves (1895–1985), *Poems 1929*, *Poems 1926–1930* (1931); W. B. Yeats (1865–1939), *The Tower* (1928), *The Winding Stair* (1929); Gerard Manley Hopkins (1844–89), *Poems*, 2nd edn, ed. Charles Williams (1930); Wilfred E. S. Owen (1893–1918), *Poems*, 2nd edn, ed. Edmund Blunden (1931).

[3] 'Nor was that final, for about that time' (Oct. 1927): *The English Auden*, ed. Edward Mendelson (1977), 24.

[4] See Edith Sitwell (1887–1964), Introduction to her anthology *The Pleasures of Poetry: First Series* (1930), in which she praises and appraises poetic effects of alliteration, assonance, and dissonance at length (75 pages).

[5] See D. H. Lawrence (1885–1930), *The Man Who Died* (March 1931), the Resurrection story sexualized; Eliot, *The Waste Land* (1922), formed from fragments of a vain Grail tradition; the poet E. E. Cummings (1894–1962), turning from poetry to a prose dream play *Him* (1927) and drawings and paintings *CIOPW* (Jan. 1931).

Time passes and now is other,
Is knowledge in him now of other,
Cries in cold air, himself no friend.
In grown man also, may see in face
In his day-thinking and in his night-thinking
Is coarseness and is fear of other,
Alone in flesh, himself no friend.[6]

The Charade which takes up half the book[7] is, as a whole, his most powerful work (what is not so easy in a short poem); one knows where one is before the end. And where one is (usually, in Auden) is in a world without miracles, which works out always in an equation or bathos. Much modern poetry presents bathos but not mere bathos; for a bathos which still comprehends the dignity of its precedents is thereby raised to tragedy.

And one can un-kernel bathos from what is prima-facie triumph. Of such triumphs modern poetry is a criticism—it discloses a peripeteia not in time but there from the beginning (there automatically); the stock values are found facile:

It is seen how excellent hands have turned to commonness,
One staring too long, went blind in a tower,
One sold all his manors to fight, broke through, and faltered.[8]

Just as art can make of suicide something vital, so here is the attempt to make of these falterings a rhythm. Hence he should certainly be read by those of us who are not old enough to go, or think we go, straight forward.

Reply to 'An Enquiry' in *New Verse*[1]

An Enquiry

1. Do you intend your poetry to be useful to yourself or others?
2. Do you think there can now be a use for narrative poetry?

[6] 'Coming out of me living is always thinking' (May 1929): *The English Auden*, 38.
[7] *Paid on Both Sides* (1928): ibid., 1–17.
[8] 'Taller to-day, we remember similar evenings' (March 1928): ibid., 26.

[1] *New Verse*, no. 11 (Oct. 1934), 2 and 7. 'An Enquiry' was sent to a number of poets by the editor Geoffrey Grigson. W. H. Auden, C. Day-Lewis, Stephen Spender did not reply; among the poets who did reply were: Laura Riding, Robert Graves,

3. Do you wait for a spontaneous impulse before writing a poem; if so, is this impulse verbal or visual?
4. Have you been influenced by Freud and how do you regard him?
5. Do you take your stand with any political or politico-economic party or creed?
6. As a poet what distinguishes you, do you think, from an ordinary man?

MacNeice's Reply

1. Mainly to myself; but I find it a very helpful detour to try to make my poems intelligible and interesting to others.
2. Yes. Narrative poetry should, logically, supersede the novel.
3. Sometimes a spontaneous impulse (its nature varies); often I have a vague feeling of deficiency which I try to fill out with a poem; this first deliberate and tentative poem is often followed quickly by a second poem which shapes itself and is usually better than No. 1.
4. Not to my knowledge. I feel that most of those artists who use him as a sanction misrepresent him. He will be more helpful when they reorientate him.
5. No. In weaker moments I wish I could.
6. Dissatisfaction with accepted formulas. But most of the time one is not a poet and is perfectly satisfied.

Modern Writers and Beliefs[1]

The Destructive Element. By Stephen Spender (Cape. 8*s.* 6*d.*)

This book is a virile, delicate, and, above all, serious piece of criticism. Mr I. A. Richards, writing of *The Waste Land*, quoted Conrad—'In the destructive element immerse. That is the way'.[2] Mr Spender's study of the destructive element begins

Wyndham Lewis, Dylan Thomas, Herbert Read, E. E. Cummings, Conrad Aiken, Norman Cameron, Wallace Stevens, William Carlos Williams, Marianne Moore, Edwin Muir, Allen Tate.

[1] *Listener* 13: 330 (8 May 1935), suppl. xiv.
[2] Stein's advice to Jim in Joseph Conrad's *Lord Jim* (1900), chap. 20, 1923 edn, 214, is quoted by I. A. Richards, *Science and Poetry* (1926), 65 n., in reference to T. S. Eliot's *The Waste Land*, and so by Spender, Introduction, 12.

with Henry James. He next discusses 'three individualists', Yeats, Eliot, and Lawrence. Finally, in a section entitled 'In Defence of a Political Subject', he presents very fairly the literary problems of his own generation. One might object that the thread joining these writers is an arbitrary one. Mr Spender maintains that they are all concerned with the same 'political subject'—call it x. But if Henry James was only concerned with it to paint its opposite y, and if Yeats and Eliot are only concerned with it to ignore it (*il gran rifiuto*),[3] and if Lawrence obstinately and myopically thought of it as z, one might dispute Mr Spender's right to lump these four writers in with Messrs Auden, Upward[4] and the other apostles of something positive. But, having read this book with a moderately open mind, I do not find that the thread or plot of it is arbitrary. Mr Spender uses 'political' in a very wide sense; many readers would recognize the common quality of his subjects sooner if it were called *seriousness* (or we might revive Arnold's touchstone, σπουδαιότης).[5] For no one, I trust, will deny that all these writers are more serious than, say, Arnold Bennett. And if one asks why Bennett was not serious, one might find in the end that it was because he was not politically minded.

Mr Spender admits that lyricism can flourish without politics (he has no wish to wipe out Mr de la Mare), but would deny this possibility of the novel, the drama, etc. And he is probably right. Jane Austen, in this sense, was a political novelist. We can't write like Jane Austen (alas!) because we haven't (thank God!) got her politics. And if we haven't got some kind of politics, we can't perhaps write novels or such works at all. Now Yeats and Eliot have a kind of inverted politics which they couldn't get on without, any more than the hermit in the Thebaid could get on without sex.[6] Mr Spender in this book works through what he regards as the noble perversions of his subject until, in the last section, he comes to the comparative normality of his contemporaries.

[3] 'The great refusal': Dante, *Inferno* iii. 60 (i.e. abdication of responsibility). Cf. MacNeice's *Autumn Journal* 1938 (1939), end xxi: *The Collected Poems*, ed. E. R. Dodds (1966, corr. 1979), 145.

[4] Edward F. Upward (1903–), novelist-friend of Isherwood and Auden.

[5] 'Earnestness, seriousness, goodness', called 'high seriousness' by Matthew Arnold, 'The Study of Poetry' (1880), *Essays in Criticism: Second Series* (1888).

[6] Not the Greek Thebes of Statius' *Thebaid*, but the Egyptian Thebes of the early Christian desert monks: the hermit Heron lapsed and contracted venereal disease—Palladius, the fifth-century Greek chronicles *The Lausiac History*, XXVI. iv; see also XXXVIII. xi, xiii and XLVII. v.

His view of his perverts is briefly this. James worshipped aristocracy but saw through the aristocrats and so turned to an apotheosis of dead virtues. Eliot, despairing of the Church on earth (?), falls back like James on another world, a death-world.[7] Yeats falls back on an esoteric blend of aristocracy and magic. Mr Spender is extremely interesting on James and refreshingly acute on Eliot, whom he convicts of a fascist strain of thinking and of some other blind spots. On Yeats he is less adequate; perhaps takes his algebra too seriously.[8] Yeats is predominantly aristocratic, but there is an anti-Yeats in him which would repay study.

Over against these three writers he puts Lawrence, but Lawrence was also an individualist. He was not, however, a reactionary, an escapist. But he cried in the wilderness and a wilderness his work remains. Mr Spender quotes a passage from a letter which, as he says, heralds something which Lawrence did not himself fulfil.[9] It is the same thing that the Communist writers are now fumbling after, some of them without knowing it.

The splendid thing about Mr Spender is that he knows it. Communism is for him not a mere economic reshuffle or an inverted individualism. The individualist is an atom thinking about himself (Thank God I am not as other men); the communist, too often, is an atom having ecstasies of self-denial (Thank God I am one in a crowd); and this too is attitudinizing. It is essential to get rid of this atomist conception of personality, which psychology has undermined from below and which true communism ignores from above. The ego as an indestructible substrate is as obsolete as the old philosophical conception of 'substance'. Yeats has recognized this in insisting that there are no hard and fast, no private minds. Communism in the truer sense is an effort to think, and think into action, human society as an organism (*not* a machine, which is too static a metaphor).

[7] But see T. S. Eliot, 'The Hippopotamus' and 'Mr Eliot's Sunday Morning Service' (both 1919–20) for negative-ironic views of the Church. Eliot's 'death-world' appears notably in 'The Hollow Men' (1925).

[8] Yeats's 'algebra' or quasi-mystical system is found in *A Vision* (1925; 1937).

[9] D. H. Lawrence's famous letter to Edward Garnett, 5 June 1914, defended *The Wedding Ring* (draft of *The Rainbow*): 'There is another ego, according to whose action the individual is unrecognisable, and passes through, as it were allotropic states . . . of the same single radically-unchanged element'—*The Letters of D. H. Lawrence*, 5 vols., ed. J. T. Boulton and G. J. Zytaruk, ii (1980), 182–4.

Mr Spender looks forward to a synthesis (it might better be called a dialectic) of communist thinking with psychological thinking. Psychology will be a check on ideology. At the moment, even the most intelligent communist tends to relapse into crude generalizations. Thus Mr Spender quotes, apparently with approval, Lenin's statement 'Art belongs to the people. It ought to extend with deep roots into the very thick of the broad toiling masses. It ought to be intelligible to these masses and loved by them, etc.'[10] Lenin is here repeating the fallacies of Tolstoy.[11] We need only ask—If this applies to art, what about higher mathematics or metaphysics? Are the mathematician and the metaphysician to be limited by the Highest Common Factor of the masses' understanding? Mr Spender, fortunately, has far too sensitive a mind for these categorical imperatives. That is why he has written a book which means something.

Translating Aeschylus[1]

Aeschylus: *The Seven Against Thebes*. Translated by Gilbert Murray. (Allen and Unwin. 3*s*.)

Professor Murray is our leading Hellenist and no one would impugn either his scholarship or his enthusiasm. But as a verse-translator of the Greek dramatists he is, though readable, neither a good translator nor a good poet. That he is readable is shown by his sales. He has now translated eight plays of Euripides, six of Aeschylus, the *Oedipus Rex* of Sophocles, and the *Frogs* of Aristophanes (the last rather emasculated).[2] Most of these

[10] Spender, *The Destructive Element* (1935), chap. 13, 229–30, is quoting Lenin's commentary on art made to Klara Zetkin from Max Eastman, *Artists in Uniform* (1934), 222, in Eastman's own translation from German, *not* taken from the current English translation of Zetkin's *Reminiscences of Lenin* (1929), 14.

[11] Leo Tolstoy, in *What is Art?* (1898), turned against his earlier work, claiming that the highest (religious) art must 'infect' people with feelings of love of God and man, so ruling out some works of Shakespeare and his own fiction as 'bad art'.

[1] *Spectator* 154: 5576 (10 May 1935), 794. MacNeice was translating Aeschylus' *Agamemnon* 1935–6 (published Oct. 1936; performed 1 and 8 Nov. 1936 by the Group Theatre).

[2] See publisher's announcement opposite title-page of Murray's translation. G. Gilbert Murray (1866–1957) was Regius Professor of Greek at Oxford 1908–36, predecessor in this position to E. R. Dodds, MacNeice's Birmingham mentor.

translations have run into their tens of thousands, though those of Aeschylus have not sold so well.

What do people want in an English version of Aeschylus, if they want it at all? Those who have a little Greek will most likely want a fairly exact crib; this Professor Murray does not provide. Those who have no Greek will want either a version which 'puts across' the original or something which will stand on its own feet as a work in English. Professor Murray, I take it, wishes to put across his original; he takes liberties not for their own sake but in order to save qualities in his original which he regards as more important than word for word accuracy. Dryden said 'A Noble Author would not be pursu'd too close by a Translator. We lose his Spirit, when we think to take his Body.'[3] It may be maintained that neither Dryden nor Pope really took the spirit of their originals, but one thing which they certainly had in common with them was technical virility.[4] And this is one thing which Professor Murray lacks. His diction is weak, though on the whole (excepting occasional 'withals' and 'what times') pure and dignified. The dialogue he does, as usual, into rhyming couplets of the William Morris type, a couplet which has the peculiarity of being a looser (and blanker) medium than blank verse.[5] In choosing metres for the choruses he tries to vary them appropriately to the variety of the Greek. Thus in this play he hurries his English to represent the hurried Dochmii of the first chorus, while he represents as follows the Ionic rhythms of the chorus beginning *πέφρικα τὰν ὠλεσίοικον θεόν*:[6]

[3] John Dryden, *Discourse concerning the Original and Progress of Satire* (1693), 6th last para.

[4] Masculine or strong couplets (as well as triplets by Dryden) are among the merits of these translations: Dryden's translations of Juvenal and Persius (1693), Virgil (1697), *Fables* (1700); Alexander Pope, translations of Homer (1715–20, 1725–6).

[5] MacNeice is probably thinking of loose pentameter couplets as often used by Murray and by William Morris (1834–96) in his multi-volume *Earthly Paradise* (1868–70)—not the longer couplets in Morris's translations (Virgil, 1875; Homer, 1887). Morris, like Murray an imitator of Swinburne's rhythms, was a notable predecessor of MacNeice at Marlborough College.

[6] This phrase opens the second choral ode. There are two basic forms of dochmiacs: ⏑ — — ⏑ — and — ⏑ ⏑ — ⏑ —, according to Christopher Dawson's trans. with commentary of Aeschylus' *The Seven against Thebes* (1970), 38; and two basic kinds of ionics: — — ⏑ ⏑ and ⏑ ⏑ — —.

I have fear of One that Watcheth till the House of Kings be broken;
'Tis a Word, or 'tis a God, but unlike the Gods we know;
Ever true and ever evil, in a father's fury spoken
And for Oedipus fulfilling to the end his will of woe.
The last meeting of his children draweth near;
Death is here.[7]

This chorus shows the weaknesses of Professor Murray's method. A literal translation line for line would be:

1 2 3 1 2 3 1
'I shudder at the house-destroying | god, to gods unlike, | all-truthful
2 3 2 1 1 2
evil-prophet, | Erinys invoked by a father | to bring about those over-
3 1 3 2 1 2 3
angry | curses of crazy Oedipus: | A child-destroying quarrel is this
4
that spurs them forward.'

(The numbers denote the order of the words in Greek.)

(a) In the first line Professor Murray so dissolves the very definite phrase τὰν ὠλεσίοικον θεόν[8] that we do not know *who* is going to break the House of Kings.

(b) ''Tis a Word' represents Ἐρινὺν,[9] but the uninstructed reader will not gather this.

(c) In the last line ἔρις[10] becomes a 'meeting' (it is not clear that it is not a friendly one), and the strong verb ὀτρύνει[11] is merely melted away.

(d) The English capital letters, 'House of Kings', etc., seem to me to make it all vaguer. So much for slavish details. To my taste the literal version given above seems, even as a whole, to be both better English and more Aeschylean than Professor Murray's. Write the former in lines and it will read as tolerable free verse. Not that I would recommend a free verse translation, but I think a translation should start from the Greek, preferably

[7] Murray's *Seven Against Thebes* (1935), 61.
[8] Literally, as MacNeice puts it above, 'at the house-destroying god(dess)'.
[9] *Erinys*, or Fury, an avenging deity.
[10] *Eris*, 'strife' or 'quarrel'.
[11] 'Rouses, presses, urges on'.

line for line. Diction and rhythm will then differentiate. A touch of Gerard Manley Hopkins might have helped Professor Murray. Thus if for 'Hark! in the gates the bronzen targes groan'[12] we substitute 'Hark! in the gates the bronze shiélds gróan,'[13] we improve both rhythm and diction and so make the whole more real. This is perhaps where the non-scholar may translate better than the scholar. His Greek original is so real to a scholar like Professor Murray that it is probably never out of his mind, and so he cannot see what the English looks like just as English.

Poetry To-day[1]

Poets do not know (exactly) what they are doing, for if they did, there would be no need to do it. So much of truth is there in the Plato–Shelley doctrine of Poetic Inspiration. Poetry is not a science and it is more than a craft. This is why, when the poet tries to explain his work, he is much less helpful than the mechanic explaining an engine. But it is a human characteristic that the poet must try to explain and the reader to comprehend why, how and what the poet writes. What both of them want, or at least what they get, is a collection of working hypotheses, often mutually contradictory. Both writer and reader will lose heart unless they can put their finger on something apparently definite, in motive, method or end achieved. Thus even the psychologist having giddied you into a glimpse of chaos will promptly cover it up with an hypothesis.

People will not read poetry unless they think they know what they are going to get from it, and people will not write poetry unless they think they know what they are driving at. Hence the importance of poetic theory and criticism. But what is needed is not necessarily the best criticism. The best criticism, like the best philosophy, tends to be negative. It is inferior philosophy,

[12] Murray's *Seven Against Thebes*, 34.

[13] Five stresses in an octosyllabic line, three stresses springing together at the end, very like Hopkins's sprung rhythm.

[1] *The Arts To-Day*, ed. Geoffrey Grigson (6 Sept. 1935), 25–67. Contributors, besides MacNeice, included: W. H. Auden ('Psychology and Art To-day'), Grigson ('Painting and Sculpture To-day'), Arthur Calder-Marshall ('Fiction To-day'), and John Grierson ('The Cinema To-day'). See E. B. Osborn's review in the *Morning Post*, 10 Sept. 1935, 14, praising MacNeice's article as 'the most brilliant piece of writing in the book'.

like pragmatism, that influences action, and it is a narrow and limited criticism that encourages the production of art. What the artist and the reader need is an Aristotle or a Dr Johnson.

Poetry at the moment is becoming narrower and less esoteric. The narrower it becomes, the wider the public it represents and the nearer it comes to being popular. It will not become so narrow as to be truly popular (representing the masses) but it seems at the moment in England to be reaching a stage where it will represent and be acceptable to a considerable minority. Very few poets can dispense with a public and be content, like Blake, to have their works printed in eternity,[2] it is therefore only common sense to acquiesce in the present trend of English poetry. Before trying to assess this present trend as exemplified in the new school of young English poets, I shall consider the development of modern verse by their immediate predecessors. But first of all I will dogmatize.

In every field of æsthetics people have been searching for what is *essential*. These investigations are mostly cant and clamour. You may not eat the shell of a nut but you can't grow nuts without shells. Artistic organisms are too inextricably complex to be amenable to deliberate vivisection. You cannot divorce the substance of a poem from its accidents. No amount of theorizing will give you the essential in a bottle. What then are we to do? Must we renounce theory altogether? To banish theory is as much of a half-truth and a whole lie as to make theory omnipotent. The functions of theory are propædeutic, prophylactic, and corrective; just as in learning to play tennis. When it comes to the point, the work is done with the hands. I hope it will not seem precious to speak of the hands of the soul. Poetic Inspiration seems to me as well symbolized by the psychic hands of the individual poet himself as by any annunciation *ab extra*.

The theorist has, for example, always divided Form and Content. The hands of the poet know better. This crude distinction has had its uses but it has made the writing of poetry appear superficially easy but actually impossible: all you have got to do is get the Right Content and put it into the Right Form. This would take to eternity. Take rhythm for example, which, in speaking of poetry, we put on the side of Form. But any particular

[2] William Blake, letter, 21 Sept. 1800, to John Flaxman: *Complete Writings with Variant Readings*, ed. Geoffrey Keynes (1966), 801–2.

rhythm x is always the same—blank x, until the so-called Content comes and *enforms* it. One of the good deeds of modern thought is to have remembered the abstract nature of abstractions (see e.g. Gentile, *Mind as Pure Act*).[3] We no longer spend our time trying to jigsaw together a preconceived Form and Content. And so good-bye to the *mot juste* ; on a correspondence theory of art all the *mots justes* would have been used up long ago. Language on one side and thought on the other are dead weights on that kind of theory. The history of recent poetry is a history of various reactions against dead weights.

The common factor in these reactions appeared to be a revolt against tradition. This was not so. All the experimenting poets turned their backs on mummified and theorized tradition, but the more intelligent realized that living tradition is essential to all art, is one of the poles. A poem, to be recognizable, must be traditional; but to be worth recognizing, it must be something new. Some poems are merely new at a first reading and we do not like them afterwards; others we like merely for their familiarity (e.g. those we have heard as children). It is snobbish to condemn these two classes of poems, but the higher poem should be partly familiar from the first, and new at the latter end. Whatever the 'true function' of poetry is, there is something idolatrous or fetishistic in our pleasure in it. One can make an idol, like a fashion, out of anything; it is part of our Self-Respect to fit up a good idol. And here again we must beware of fixing a great gulf of abstraction between the Self and the Object (or the Product).

The poem, like the idol, is a kind of Alter Ego. But the Alter Ego is another polarized concept. As Ego it is self-expression; as Alter it is escape from Self. Hence the dangers in explaining a poem through its author. Hence also the falsity of the hackneyed distinction between Poetry of Escape and poetry which represents or interprets 'Life'. And the generalizations so often offered by poets themselves we must accept only as dramatic; they are not scientific rules but are merely a moment in the context of the poet's life and work, which can be contradicted by other moments in his life and work, just as one character in a play can give the lie

[3] Giovanni Gentile, *The Theory of Mind as Pure Act*, trans. H. W. Carr (1922), chap. 8, paras. 1–5, 96–100.

to another character. '*Je rature le vif*'[4] is not the same kind of statement as $2+2=4$ (or, to be safer, as $0=0$). Posterity affects to put dead poets and movements in their place; to tell us their real significance and cancel out their irrelevances. This habitual procedure of posterity is, like other affectations, useful in that it is tidy and saves thinking (I do not mean one-way thinking). Most people can only afford the time to see contradictions in their contemporaries. The self-contradictory is what is alive, therefore for most people the most living art is contemporary art. Yet people are ungrateful; they prefer the dead to the living and try to kill even their contemporaries by looking a hundred years forward. I continually hear people saying 'Yes, but I wonder what people will say of him a hundred years hence', or 'I dare say, all the same, posterity will think more of Mr X.' They herein miss the point. If we do our duty by the present moment, posterity can look after itself. To try to anticipate the future is to make the present past; whereas it should already be on our conscience that we have made the past past. We fail to appreciate a great poet like Horace because we don't let him puzzle us; to indicate the concreteness of Horace, as of Mr Yeats, we should need a dialectic of opposites. If poetic criticism is to develop, it must give up one-way thinking. As it is, the man who reads a poem and likes it, is doing something far too subtle for criticism.

I will not try to explain further why it is necessary for some people to write, and for more people to read modern poetry. Assuming that this is so, it is important that the individuals of both these classes should co-operate; to be mutually intelligible these poets and readers must have a more or less similar orientation of tradition. They must avoid the two extremes of psittacism[5] and aphasia. There are cultured people in England to-day who

[4] Literally, 'I erase what's alive', meaning roughly, 'Of the lives written by life I cross out whatever I like or don't like': Paul Valéry, 'La Soirée avec Monsieur Teste' (1896), *Oeuvres*, Éditions Gallimard, 2 vols. (1957, 1960), ii, 17: 'Voici ses [de M. Teste] propres paroles: "Il y a vingt ans que je n'ai plus de livres. J'ai brûlé mes papiers aussi. Je rature le vif . . . Je retiens ce que je veux. Mais le difficile n'est pas là. *Il est de retenir ce dont je voudrai demain!* . . . J'ai cherché un crible ['sieve'] machinal . . ." '.

[5] For psittacism, mimic mockery, and the parrot, see MacNeice's juvenile poem quoted in his *Modern Poetry* (1938), 39–40, his play *Out of the Picture* (1937), his late poem *Autumn Sequel* 1953 (1954), 'Budgie' (1962) in *The Burning Perch* (1963), and elsewhere. Cf. John Skelton's 'Speke, Parrot' (1521); cf. also 'psittacosis' in MacNeice's Oxford essay 'We are the Old' (1930) in *Modern Poetry*, 73.

write poems which are mere and sheer Shelley; these are psittacists; they are betraying themselves (and, incidentally, betraying Shelley). There are also the enthusiasts (mostly Americans in Paris) who set out to scrap tradition from A to Z; this should logically lead to aphasia; that they do not become quite aphasic is due to their powers of self-deception. How are we to do justice, not to the segregated Past or Present, but to their concrete antinomy?

The problem is especially difficult for us because, unlike our more parochial predecessors, we have so many Pasts and Presents to choose from. We have too much choice and not enough brute limitations. The eclectic is usually impotent; the alternative to eclecticism is clique-literature. The best poets of to-day belong to, and write for, cliques. The cliques, lately, have not been purely literary; they identify themselves with economic, political or philosophical movements. This identification is more fruitful when it is voluntary; I am told that Communism in Russia and Fascism in Italy have not, as yet, elicited much good poetry to order. The poet must be primarily a poet and this is still possible in England. But the common assumption that English poets have always been free lances is a gross misrepresentation. Those who admire the 'freedom' of the free lance should take a course of Spinoza;[6] the best English poets have been those most successfully determined by their context. The context must be a suitable one. The English context is now more congenial to poets than it has been for a long time.

English poetry of the nineteenth century was doomed by its own pretentiousness ('poets are the unacknowledged legislators of mankind').[7] Victorian scepticism made us draw in our horns: 'Poetry is a Criticism of Life.'[8] But Samuel Butler and his kind were the Critics of Life; the English poets of the 'nineties, misapplying some recent hints from the French, left Life and Mankind out of it and turned to cultivating their gardens. A suburban individualism prevailed, the penalty for the bumptious

[6] Benedict de Spinoza (1632–77), deterministic philosopher of a self-enclosed, deductive system, lived a tragic life as maker of lenses quite unrelated to his lofty philosophy.

[7] 'Poets are the unacknowledged legislators of the World': final statement of P. B. Shelley's *A Defence of Poetry* 1821 (1840). MacNeice's wording echoes that of T. S. Eliot's *The Use of Poetry and the Use of Criticism* (1933), 25.

[8] Matthew Arnold, 'The Study of Poetry', para. 2.

anarchism of the Romantic Revival. The poets of the 'nineties and the Georgians[9] who succeeded them were crippled by a reaction from the prophets; they did not dare to be moral, didactic, propangandist or even intellectual; fear of being thought hypocritical precluded them from interest either in God or their neighbour. This bogey of hypocrisy had hamstrung our intellects.

Contemporary with this fairly homogeneous suburban movement (which began with æstheticism and ended with a castrated nature-poetry and occasional pieces: see the *Georgian Anthologies*, *passim*) there were certain sturdier freak-growths, e.g., Mr Kipling's jingoism and Mr Housman's pastorals. Neither Kipling nor Housman has had successors, though they both anticipated certain freedoms of diction and Housman anticipated the all-pervading irony of post-War poetry. More important was the Irish movement, where poetry was healthily mixed up with politics. Yeats's early poems, which many would take as typical escape-poetry, were very much more adulterated with life than e.g. the beery puerilities of Messrs Chesterton and Belloc.[10] We now laugh at the Celtic Twilight and at the self-importance of these dilettante nationalists, but their *naïveté* and affectation had manured the ground for poetry. Where it is possible to be a hypocrite, it is also possible to be a hero, a saint, or an artist. It was hardly possible for a poet to be a hypocrite in England in the pre-War period. Hence the thrill (and subsequent, as it seems to us, hypocrisy) of writers like Rupert Brooke, when the War broke out.

We must not too readily assign the War as a cause of developments in the arts. By the time the War broke out, Mr Pound and his Imagists had already asserted themselves, Mr Eliot had read his Laforgue,[11] Mr Yeats was working steadily to make his verse less 'poetic', Free Verse was an old story and Marinetti had invented Futurism.[12] But in England at any rate this left-wing literature did not become notorious till after the War. And in 1922 appeared the classic English test-pieces of modern prose and

[9] Five anthologies of *Georgian Poetry 1911–1922*, ed. Sir Edward Marsh, included mainly pastoral verse by such poets as Rupert Brooke, W. H. Davies, James Elroy Flecker, Walter de la Mare, as well as some First World War poetry. Cf. *Modern Poetry*, 8–9, 11.

[10] G. K. Chesterton (1874–1936); Hilaire Belloc (1870–1953).

[11] Jules Laforgue (1860–87), romantic ironist in verse that had direct influence on T. S. Eliot in *Prufrock and Other Observations* (1917).

[12] F. P. Marinetti (1876–1944), poet, founder of futurism 1909–16.

verse—*Ulysses* by James Joyce and *The Waste Land* by T. S. Eliot. To most of the intelligent minority both these works appeared incoherent and obscure; *Ulysses* was also considered overwhelmingly obscene. The same minority would now agree that neither work is so difficult if approached from the right angle, and that as for obscenity there is far more in any popular magazine. We have new standards of coherence and of poetic meaning. Joyce as well as Eliot has had great influence on our poetry—for two reasons. As a technician he uses words in that subtle way which is usually the privilege of poets. As a very sensitive observer his acceptances and rejections (vulgarly called his realism) have sanctioned the acceptance or rejection of certain subject matters in poetry.

Joyce and Eliot were accepted as protagonists of the New Order. Both these writers have an aristocratic objectivity or impersonality due to a wide acquaintance with European culture. The hot-gospelling element for post-War England was supplied by D. H. Lawrence. The young contrived to be influenced by these three very different writers simultanaeously. The three had this in common, that they transgressed the limits of the pre-War Suburban Individualists. Eliot reintroduced into poetry first the intellect and later Christianity; Lawrence had the welcome bad taste to be a prophet; Joyce, instead of cultivating his garden, attempted with superb effrontery and industry to assimilate the modern world.

For a decade after the War these three were dominant. But Lawrence was too vague and Eliot and Joyce were too wide. Before discussing the present very important poetic reaction, I will mention some of the experiments tried in English poetry during this post-War period. In France[13] the *Surréalistes* appeared soon after the War, though knowledge of their work is beginning only now to penetrate into England; Paul Valéry, on the other hand, broke a fifteen years' silence in 1917 with *La Jeune Parque*. English poets have, however, been mainly (and too much so) influenced by other English poets; the imitators of Eliot would have done well to read Eliot's originals, just as the irrationalists would have done well to read their Freud. On the whole the period

[13] MacNeice's note: 'I make no attempt in this article to assess contemporary poetic movements in foreign European countries, or even in America. It goes without saying that all the arts are more cosmopolitan than they were, but my scope is purposely narrow.'

is too literary. It is only now that poets are learning to forget their literature. Here Mr Pound is one of the worst offenders. For very many years he has been repeating, rather hysterically, that he is an expert and a specialist; but he has specialized his poems into museum pieces. His recently published *A Draft of Thirty Cantos*[14] is, I suppose, a good piece of its kind; like *Ulysses* it is certainly far the longest of its kind. I maintain, however, that the kind is not one to be encouraged. What Mr Grigson has called 'the cultural reference rock-jumping style',[15] even if feasible in a poem of the length of *The Waste Land* where every reference can be manœuvred to pull its weight, is bound to lose its virility in a work as vast as the Cantos. Apart from this, Mr Pound's effects are very monotonous; he uses the same cadences again and again for glamour, and the same contrasts again and again for brutality (or reality). The faults in his work should remind us of certain practical, if pedantic, truths. Quantity must always affect quality. A metre of green, as Gauguin said, is more green than a centimetre,[16] but a bucket of Benedictine is hardly Benedictine. Mr Pound does not know when to stop; he is a born strummer. The second Aristotelian truth against which he offends is this. The poet's method must be specifically poetic (something between the philosopher's and the somnambulist's). Or we may put it that poetry is a genre somewhere between play and science. The poet must ape neither the scientist nor the child with his ball; Mr Pound apes them both. The child, and the ordinary man when he is being childish, exploit an activity which is other than the poetic. They go out of their way to use a home-made jargon, nicknames, private allusions, tags of special knowledge. It can, of course, be maintained that all this is poetic

[14] *A Draft of XXX Cantos* (Paris, 1930; New York and London, 1933). Cf. *Modern Poetry*, 163–4.

[15] In reviewing Ronald Bottrall's second poetry book *Festivals of Fire* (1934), Geoffrey Grigson remarked, 'Mr. Bottrall imitates the cultural-reference-rock-jumping style of Eliot and Pound': 'Two Poets', *New Verse*, no. 8 (April 1934), 19.

[16] Paul Gauguin (1848–1903) quoting Paul Cézanne (1839–1906): '. . . Cézanne dit avec un accent méridional: "Un kilo de vert est plus vert qu'un demi-kilo." Tous de rire: il est fou!'—Paul Gauguin, 'Diverses choses', *Oviri: Écrits d'un sauvage*, ed. Daniel Guérin (1974), 174; reiterated 177. 'Cézanne says, with his accent from the Midi: "A kilo of green is greener than half a kilo." Everyone laughs: he's crazy!'—Paul Gaugin, 'Miscellaneous Things', *The Writings of a Savage*, trans. Eleanor Levieux, ed. Daniel Guérin (1974), 141, reiterated 143–4. Cited again by MacNeice in *Zoo* (1938), 110.

or artistic efflorescence, that when a middle-class Englishman quotes French he is satisfying an artistic impulse. It can similarly be maintained that dreams, alcohol, sex or looking at the landscape give us a pleasure (or a release or a consummation or anything else) identical with that given us by poetry, and that we take to reading poetry for the same reason as we take to these other things. Thus Mr E. E. Cummings says that his poems are competing with roses and locomotives.[17] All this may be so, but if poets are to continue to exist (as they will so long as there are people not contented with roses and locomotives) it is essential that the poet should do more than give you a drink or tell you to look at the view; he must use words and at that he must still give you more than the child gives you when he distorts or jingles for his pleasure or the grown man when he makes a pun or a quotation. We may agree with the psychologist who tells us that all these activities are fundamentally akin. Mature and civilized man is concerned with the surface. We must maintain differences. Or else we must be honest monist-nihilists and not meddle with the arts. It is honest to say of a picture 'It doesn't matter which colour goes there, it's all canvas underneath,' but it is not honest, if one holds this point of view, to become a painter. There has been a great deal recently of this (often unwitting or half-witted) dishonesty; see the past numbers of *Transition*, *passim*, or Mr Pound's recent *Active Anthology* or even the earlier, almost forgotten *Imagists*.[18] It is the kind of dishonesty which was magnificently attacked by Mr Wyndham Lewis in *The Enemy No. 1*.[19] I am summarizing the 'Enemy's' attack when I say succinctly that Mr Pound lacks grip; professing to offer us poetry

[17] Foreword to *is 5* (1926): E. E. Cummings, *Complete Poems* (1968), 223.

[18] Cf. *Modern Poetry*, 160. Eugene and Maria Jolas, founders of the review *Transition* (1927–38), tried to conduct through it a 'revolution of the word' using as principal text Joyce's 'Work in Progress' *Finnegans Wake* (1939): see Richard Ellmann, *James Joyce*, rev. edn (1982), 587–9. Ezra Pound, also a literary programmist, edited such poetic anthologies as *Des Imagistes* (1914) and *Active Anthology* (1933) in attempting radically to alter the writing of poetry: *Des Imagistes* included poems by Hilda Doolittle ('H.D.'), Richard Aldington, Ezra Pound, F. S. Flint; *Active Anthology*, poems by William Carlos Williams, E. E. Cummings, Marianne Moore, Louis Zukofsky, Basil Bunting, as well as Eliot and Pound.

[19] 'The Revolutionary Simpleton', *The Enemy*, ed. P. Wyndham Lewis, 1: 1 (Jan. 1927), 24 chaps, includes in chaps 9 and 15 an attack on Pound, and in chap. 16 some mockery of Lewis's friend Joyce as 'poet of the shabby-genteel, impoverished intellectualism of Dublin', 97.

he is always falling back on easier substitutes. These substitutes, thanks to the psychologists and the flux-philosophers, are to-day at a high premium; this is made an excuse for both excessive laziness and excessive industry. Further instructive examples of this intellectual epidemic are found in the poetry of Mr Robert Graves, Miss Laura Riding, the Sitwells and Mr E. E. Cummings.[20]

Mr Graves and Miss Riding are very conscious moderns and purists. Mr Graves began with an admiration for ballad and nursery rhyme and produced pretty little poems of that type, using a rather Georgian technique. After the War as a remedy for shell-shock he took to reading Freudian psychology and analysing the latent symbolic elements in his own work. He decided that the manifest-content of most poetry is irrelevant; his careful pruning of content, combined with technical austerities, has left his more recent verse bare and arid and approximating more and more to the metaphysical tenuities of Miss Riding, who is obsessed by the paradoxes of Nothingness.[21] They are both painfully lacking in worldly content.

The Sitwells, on the other hand, offered plenty of the traditional kind of content—'plot', 'story', 'characters' and even 'morals', thinly disguised by a kind of syncopation. They caused exaggerated surprise among their contemporaries merely because they sowed with the whole sack certain tricks which had previously been used more sparingly. Their mythology was borrowed mostly from the eighteenth century, the nursery, and the circus side of post-Impressionist painting. They have inverted the usual flux-attitude; the dominant characteristic of their work is a devaluation of the living moment and an apotheosis of memory (cp. Proust). They are topical only in order to be satirical. Their manner of dealing with the past is Pre-Raphaelite; it is not therefore surprising that they rarely attain higher values than those of the fairy-story (just as the only good that came from the Pre-Raphaelite painters was fairy book illustrations). Miss Sitwell can write very handsome lines but she is not good at architectonic. Hence her better poems

[20] Robert Graves (1895–1985); Laura Riding Jackson (1901–); Dame Edith Sitwell; Sir Sacheverell Sitwell (1897–); E. E. Cummings (1894–1962).

[21] See MacNeice's review of Laura Riding's *The Life of the Dead* in *New Verse*, no. 6 (Dec. 1933), 18–20.

are her shorter ones (e.g. *Bucolic Comedies*). *The Sleeping Beauty*[22] is, however, a very fine jazz fairy story; she has filled in all the corners with twirls and iterations and the whole is pervaded with a kindly sentimentality (she has a country-house sympathy with such types as housemaids and spinsters). In her more recent poems she has repeated herself with a bigger ration of moralizing and a more obvious derivativeness from her favourite Old Masters. Mr Sacheverell Sitwell also is derivative in his decoration, most obviously in his octosyllables which are reminiscent of *Appleton House*.[23] His longer poems, with a looser line, were more interesting experiments but unfortunately, like his sister and like Mr Pound, he gives too much merely accumulative and not significant detail. When the atmosphere is not destroyed by cultural padding he can produce an admirable poem like 'Convent Thoughts in Cadiz'.[24]

Mr Cummings is the humorist among these poets. An American like Pound and Eliot, he cannot get over the contrast between the vulgarity of the modern world and the beauty of European culture (cp. *A Draft of Thirty Cantos*, 'Burbank with a Baedeker: Bleistein with a Cigar',[25] or even the novels of Mr Sinclair Lewis). Mr Cummings is a simple-minded poet. He has two manners, his (earlier) grand or sentimental manner with use of the vocative 'thou', and his impressionist or humorous manner which Mr Graves has described as a 'taxi-and-gin shorthand'.[26] This latter is seen at its best in his volume *is 5*. The verse here is speeded up and staccato; he has made many typographical experiments—commas in the middle of words, gaps (of all sizes) horizontal between words and vertical between lines, no capital letters where expected and capital letters where not, etc. This has irritated many people but I cannot remember meeting in *is 5* any typographical oddity which I could not see a reason for. Here Cummings compares very favourably with most of the American contributors to

[22] (1923–4). Cf. *Modern Poetry*, 52.

[23] 'Upon Appleton House', Andrew Marvell's longest poem, written in octosyllabic couplets grouped in 97 octave stanzas.

[24] (1925, 1927).

[25] 'Burbank . . . Bleistein' in Eliot's *Poems* (1920) presenting two tourists in Venice had become notorious for compressing more disjointed quotations and allusions than any poem of comparable length that Eliot had published: Grover Smith, *T. S. Eliot's Poetry and Plays*, 2nd edn (1974), 51.

[26] Laura Riding and Robert Graves, *A Survey of Modernist Poetry* (1927), 245.

Transition or *The Active Anthology*. In the matter of chopping lines Cummings chops off his lines with a sense of rhythmical or dramatic fitness, whereas in a writer like Dr William Carlos Williams[27] the arrangement into lines is entirely indifferent. Cummings is the best of his group[28] but he is unsatisfactory, just as the Impressionist painters were unsatisfactory. In art you must have an a priori. The preface to *is 5* states 'If a poet is anybody, he is somebody to whom things made matter very little—somebody who is obsessed by Making. . . .'[29] The poet's business is 'ineluctable preoccupation with the Verb.'[30] But then the poet, if he is logical, will give up writing for action; we are reminded of Marinetti's walking statues.[31]

All these poets, Pound, Graves, Riding, Cummings, and the Sitwells, have been admirably adventurous and ingenious, but, as far as living tradition is concerned, they are so many blind alleys. One can make use of their discoveries but it will not be to the same ends. To find a bridge between the dominant poetry of the early nineteen-twenties and the dominant poetry of the early nineteen-thirties we have to look back again to T. S. Eliot. Eliot as a craftsman has been greatly influenced by Pound[32]—one more instance of the greater being conditioned (*but not caused by*) the less. The difference between Eliot and Pound can be seen from their likes and dislikes. They both (if

[27] (1883–1963).

[28] MacNeice's note: 'Cummings, in *is 5* at any rate, is, thanks to his ingenious codification, a better poet than e.g. Mr Richard Aldington who tries to be clear and objective and is merely dull and sloppy.' Cummings's 'group', American poets of the 1920s and 30s centred in New York, included William Carlos Williams, Marianne Moore, and Wallace Stevens.

[29] Foreword to *is 5*, op. cit., 223. [30] Ibid.

[31] Marinetti (see n. 12 above) was a futurist poet, not a sculptor. Here MacNeice is echoing Wyndham Lewis in *The Enemy* cited above: '[The futurists] were a sort of painting, carving, propaganding ballet or circus, belonging to the milanese showman, Marinetti. One of the tasks he set them was to start making statues that could open and shut their eyes, and even move their limbs and trunks about, or wag their heads. . . . [Para.] So let us say that the "life" and "motion" ideas are seen to meet in the mechanical *moving statue* of the 1914 futurist.'—*The Enemy*, 1: 1 (Jan. 1927), 165.

[32] Eliot dedicated *The Waste Land* (1922) to Pound as *il miglior fabbro*, 'the better craftsman' (Dante *Purgatorio* xxvi. 117), acknowledging help in pruning and editing his long manuscript: evidence is given in the facsimile edn of *The Waste Land*, ed. Valerie Eliot (1971), with Pound's annotations. Pound's influence on Eliot was noticed esp. by F. R. Leavis, *New Bearings in English Poetry* (1932; 2nd edn 1950), 133.

for different reasons) admire Dante but Eliot admires, while Pound detests, Dryden.[33] More people talk about Dryden than read him but there is undoubtedly a Drydenism in the air. Mr Wyndham Lewis's recent *One-Way Song* is professedly Drydenesque.[34] One of the younger poets, Mr Charles Madge, recently went so far (in an article in *New Verse*, No. 6) as to say that all poetry is *didactic*.[35] For this general movement towards clarity and rigour Mr Eliot is largely responsible; more through his criticism than his practice.

Eliot's precisely tentative essays have reminded a world deafened by catchwords in how delicate a ratio or harmony a poem consists. But he has necessarily committed himself to a catchword or two in return, the most notorious of which is 'impersonality'.[36] To understand a man's catchword you must collate it with his tastes; among those authors whom Eliot most admires are the Jacobean dramatists and Donne, Dryden and Baudelaire. Herein (excepting Dryden?) most of us agree with or follow him. But we must notice that though Eliot, having an imitative mind and ear, has details in his work similar to all four of the above, yet in the total effect of his poems he is not only different from but alien to these writers. The reason for this is partly his basic romanticism. If you compare the defeatism of 'Prufrock' or *The Waste Land* with the apparent defeatism of the *Fleurs du Mal*, you can see why, if Baudelaire is classic, Eliot is romantic. For me

> I should have been a pair of ragged claws
> Scuttling across the floors of silent seas[37]

shows the same romantic nostalgia as

[33] See Eliot's *Homage to John Dryden* (1924) and his BBC broadcasts 1931 published as *John Dryden: The Poet, the Dramatist, the Critic* (1932).

[34] Cf. *Modern Poetry*, 188. 'Envoi' to *One-Way Song* (1933) reads: 'These times require a tongue that naked goes, / Without more fuss than Dryden's or Defoe's' (ll. 11–12): *Collected Poems and Plays*, ed. Alan Munton (1979), 91.

[35] 'Surrealism for the English', *New Verse*, no. 6 (Dec. 1933), 16. The English poet Charles H. Madge (1912–   ) was to become founder of Mass Observation 1937 and sociologist on study missions to Third World countries.

[36] Eliot's 'impersonal theory of poetry' is set forth in his influential 'Tradition and the Individual Talent' (1919): *Selected Essays*, 3rd edn (1951), 17–22.

[37] Eliot, 'The Love Song of J. Alfred Prufrock', ll. 73–4; quoted in *Modern Poetry*, 84.

Thou wast not born for death, immortal Bird![38]

No amount of wit[39] can counterbalance the mood, and in Eliot's verse the mood is dominant. But if Eliot has failed to be classical himself, his influence has been towards classicism. At a time when English poetry was sagging desperately he has restored its nervous tension.

In an essay on the Metaphysical Poets (1921) Eliot issued a kind of manifesto—'A thought to Donne was an experience; it modified his sensibility. When a poet's mind is perfectly equipped for his work, it is constantly amalgamating disparate experience; the ordinary man's experience is chaotic, irregular, fragmentary. The latter falls in love, or reads Spinoza, and these two experiences have nothing to do with each other, or with the noise of the typewriter or the smell of cooking; in the mind of the poet these experiences are always forming new wholes. . . . We can only say that it appears likely that poets in our civilization, as it exists at present, must be *difficult*. Our civilization comprehends great variety and complexity, and this variety and complexity, playing upon a refined sensibility, must produce various and complex results. The poet must become more and more comprehensive, more allusive, more indirect, in order to force, to dislocate if necessary, language into his meaning.'[40] Most of the younger post-War poets accepted this very plausible statement of the probable nature of present day poetry. Writing to this pattern they produced with great ease the two salient characteristics of difficulty and intellectuality. But the difference between Eliot and his imitators was this; a thought to Eliot is, in most cases, really an experience of the first value but to most clever young men thoughts rank far below sensations. And so their poems were frigid intellectual exercises (though very good practice for undergraduates). Mr Empson's poems seem to me to be still of this type.[41]

[38] John Keats, 'Ode to a Nightingale' (1819), st. 7.

[39] MacNeice's note: 'I have not space to discuss "wit" which Eliot restored to English poetry. It is now again realized (i) that the poet has a right to be witty but (ii) that wit is sometimes dangerous, as encouraging unconscious cheating.'

[40] *Selected Essays*, 287, 289; first half quoted in *Modern Poetry*, 163.

[41] See MacNeice's review of William Empson's *Poems* in *New Verse*, no. 16 (Aug.–Sept. 1935), 17–18, and the later *Concise Encyclopedia* entry (1963), 127–8, for which see bibliography of short prose, below.

Apart from difficulty and intellectuality the most notable thing in Mr Eliot's prospectus is the importance of the empirical element (the noise of the typewriter, the smell of cooking). This has, of course, always been present in poetry, but the manipulation of it has more often been left to the poet's hands (his psychic hands) and not to his deliberating brain. The deliberating brain tends to hamper, rather than to assist, those experiences which are forming new wholes. Mere contemporaneity will be taken as a final sanction; this will lead to fake-poetry. You cannot force the empirical element; an obvious danger also in surrealism.

We may contrast with the passage quoted from Mr Eliot the Preface to *New Signatures* (1932) written by Mr Michael Roberts. *New Signatures* was an anthology of poems by Messrs W. H. Auden, Cecil Day-Lewis, John Lehmann, Stephen Spender and others; for many people it was the first intimation that there was a new movement in English poetry. Mr Roberts says in his preface, 'The solution of some too insistent problems may make it possible to write "popular" poetry again: . . . because the poet will find that he can best express his newly-found attitude in terms of a symbolism which happens to be of exceptionally wide validity. . . . The poems in this book represent a clear reaction against esoteric poetry in which it is necessary for the reader to catch each recondite allusion.'[42] Mr Roberts then brings in the criterion of *elegance* and defends the abandoning by his poets of free verse which, he says, 'offers none of the possibilities of counterpointed rhythm which has been one of the technical delights of English verse.'[43] Finally in regard to the communism of most of these poets he says 'It is . . . not surprising that some of these poets should combine a revolutionary attitude with a respect for eighteenth-century ideals. . . . This impersonality comes not from extreme detachment but from solidarity with others. It is nearer to the Greek conception of good citizenship than to the stoical austerity of recent verse.'[44] Poetry is to be 'a popular, elegant and contemporary art.'[45]

[42] 'Michael Roberts' (pseud. of William Edward Roberts, 1902–48), Preface, *New Signatures* (1932), 11–12.

[43] Ibid., 16. Counterpoint had been highlighted by theory and practice in G. M. Hopkins's *Poems*, 2nd edn, ed. Charles Williams (1930, 1933).

[44] *New Signatures*, 18–19. [45] Ibid., 20.

This was an exciting declaration, though it was not fully borne out by the poems themselves. Mr Tessimond on the one hand remained a left-wing exhibitionist and Mr Julian Bell, on the other, a too easy reactionary.[46] As for the revolutionary, or communist, attitude, it was often so facile as to appear a mere nostrum. We are now in danger of a poetry which will be judged by its party colours. Bourgeois poetry is assumed to have been found wanting; the only alternative is communist poetry. This seems to be an over-simplification. I doubt whether communist and bourgeois are exclusive alternatives in the arts and, if they are, I suspect these would-be communist poets of playing to the bourgeoisie. And I have no patience with those who think that poetry for the rest of the history of mankind will be merely a handmaid of communism. Christianity, in the time of the Fathers, made the same threats; all poetry but hymns was bogus, no one was to write anything but hymns. It is significant that it was Tolstoy, the most vehement of recent Christians, who handed over this destroying torch to communism (see his fallacious polemic *What is Art?*).[47] This intoxication with a creed is, however, a good antidote to defeatist individualism; poets could not be expected to go on writing 'Prufrocks' and *Mauberleys*. The three most interesting poets in *New Signatures* are W. H. Auden, Stephen Spender and Cecil Day-Lewis; all three in their poems are implied communists and often propagandists. Like all propagandists (cp. Shelley) they sometimes make themselves ridiculous. Auden is often saved by his technical concentration and Spender by his technical economy but Day-Lewis, who writes longer and looser works and has not much sense of humour, has committed lamentable ineptitudes while preaching for the cause. Compare his two long poem sequences, *From Feathers to Iron* and *The Magnetic Mountain*.[48] In the former the theme is personal and is maintained throughout with dignity. *The Magnetic Mountain*,

[46] A. S. J. Tessimond (1902–63); Julian Bell (1908–37), killed in the Spanish Civil War. See A. T. Tolley, *The Poetry of the Thirties* (1975), 65–71, 110–11, 117–18, 375.

[47] See MacNeice's review 'Modern Writers and Beliefs' printed above, with n. 11, for Tolstoy's *What is Art?* (1898). The Communist declaration of 'socialist realism' in 1934 began a widespread purge in Russia of modernist art and literature. A common exaltation of subject matter 'for the people' at the expense of æsthetic considerations forms a link between Tolstoy and the USSR.

[48] (1931; 1933).

which is propagandist, begins with a very fine figure (that of the Kestrel) but falls away into slipshod satire, derivative Auden, and priggishness. Both Auden and Day-Lewis (in common with so many less articulate admirers of the Soviet) suffer from an inverted jingoism reminiscent of Kipling or Newbolt:

> A blinding light . . . and the last man in.[49]

Day-Lewis is (so far) an inferior poet to Auden and Spender, perhaps because his vision is purer and more consistent. Auden has many thoroughly bourgeois tastes and Spender is a naïf who uses communism as a frame for his personal thrills. (It is significant that he admires the now unfashionable Romantic Revival). Spender is the easiest of these poets to understand (but not to appreciate) at a first reading; Auden, who to start with was very difficult, is grinding his verse into simplicity. As simplicity, which is (from one angle only) a matter of technique, is so called for by the majority of readers, I will consider in more detail the technique of these poets. Then I will come back to their content, their beliefs and emotions.

Mr Michael Roberts has given a reason why poets are ceasing to write free verse. But many intelligent people cannot see why they began to write it. The plea is sometimes made that free verse is an attempt to express the quickened and irregular tempo of modern life. This explanation is vague and bad (like most explanations which hinge on the word 'express'). If the critic viciously separates form from content, it is difficult to see that any definite form x is especially suited to expressing any definite subject matter y; consider the vagaries of the sonnet and the various (and unelegiac)

49
> There's a breathless hush in the Close to-night—
> Ten to make and the match to win—
> A bumping pitch and a blinding light,
> An hour to play and the last man in.
> And it's not for the sake of a ribboned coat,
> Or the selfish hope of a season's fame,
> But his Captain's hand on his shoulder smote—
> 'Play up! play up! and play the game!'

Sir Henry Newbolt, 'Vitaï Lampada', st. 1, *Poems: New and Old*, 2nd edn (1919), 95, referring to Lucretius ii. 79: *quasi cursores, vitaï lampada tradunt*, 'like runners, they hand on the torch of life'. Auden and Day-Lewis are being criticized for schoolboy balladry of a propagandist kind like that of the older imperialist verse they themselves were reacting against—hence, 'inverted jingoism'.

uses of the Greek and Latin elegiac.[50] I would lodge the same objection against the doctrine of phonetic significance.[51] There are certain roughly identifiable onomatopœic noises but in the practice of nearly all poets sound has been kept in gear with meaning and it is impossible to say that the sound is significant when abstracted from that meaning. The reason for changes in technique is not to be found so simply in an 'expression' doctrine of progress. I would rather be very humble and say that changes in poetic technique are on a level with the changes each season of woman's fashions in dress. In the latter it is admitted that there is a certain deference to utility (appropriateness to the occasion) but the great majority of innovations are due to the love of novelty. And the love of novelty and variety is psychologically the mainspring of the universe.[52] Feminine fashions in dress are notoriously fickle merely because women wear clothes all the time and the fashions soon become stale. If we read poetry sixteen hours a day it would become stale sooner than it does. In poetry, of course, we keep in with the old fashions while we practise the new ones; and the old ones owe their freshness to the freshness of the new ones (e.g., it is difficult to imagine that when the Romans had written like Virgil for four hundred years, they were not sick of Virgil). Poets, therefore, began to write free verse merely because, at the time, there seemed to be nothing else to write. All the traditional forms appeared to be played out. I will take as the most obvious example the traditional line used in blank verse and in the rhyming couplet. The traditionalist sometimes maintains that this line served our purposes for four hundred years and that therefore there is no reason to discard it. A little attention will prove that the line during those four hundred years was not the same line. Marlowe, Shakespeare, Webster, Milton, Pope, Keats and Tennyson all used it differently. The differences were sometimes rhythmical, sometimes textural, but by the end of the Victorian period it seemed impossible to push it any further

[50] The classical 'elegiac'—a couplet of dactylic hexameter followed by a pentameter—is commonly confused with the classical 'elegy' composed in elegiac couplets on one of a variety of subjects: war, death, love, and similar themes.

[51] Phonetic significance or onomatopœia in poetry was being extolled by Edith Sitwell: Introduction to her anthology *The Pleasures of Poetry*, 3 series (1930–2); *Aspects of Modern Poetry* (1934), esp. chap. 6 ('Sacheverell Sitwell').

[52] MacNeice's note: 'But I do not say this frivolously. A new poem remains a higher thing than a new hat.'

without snapping its up to then elastic identity. Also with the Victorians (e.g., Tennyson and Morris) it had become an effeminate or autumnal thing. Swinburne by a *tour de force* managed to give it some virility. He did this by sacrificing its capacity for conversation (or sober narrative) to its incantational qualities; e.g.:

> Men cast their heads back, seeing against the sun
> Blaze the armed man carven on his shield, and hear
> The laughter of little bells along the brace
> Ring, as birds singing or flutes blown, and watch,
> High up, the cloven shadows of either plume
> Divide the bright light of the brass, and make
> His helmet as a windy and wintering moon
> Seen through blown cloud and plume-like drift, when ships
> Drive, and men strive with all the sea, and oars
> Break, and the beaks dip under, drinking death—[53]

This (first published 1865) is magnificent but it is a cul-de-sac, as Swinburne found out himself. At the same time Browning was pushing the blank verse line to the other (the conversational) extreme; there were perhaps a few more possibilities here and he has had more influence, e.g. on Pound and Yeats and through them on others. But the time came when most poets had to renounce nominal adherence to this tradition.

One of the results was that long poems went out of fashion. The poets of the 'nineties and the Georgians kept to small patterns with only slight technical innovations. No parallel[54] appeared to the verse of the French *Symbolistes*, perhaps because Shakespeare had already drained those wells in England. Then, before the War, the Imagists and others began writing free verse. This was verse in lines of uncertain length, without definite metre or rhyme. Rhythm is, of course, almost impossible to exclude but the rhythms were purposely those of prose or of everyday talk. This apparently random outbreak released poets from the suffocation of the Victorian salon. Next (inevitably) we find them looking back to earlier points in the English tradition. Thus in Eliot, who

[53] A. C. Swinburne, *Atalanta in Calydon* (1865), ll. 262–71.

[54] MacNeice's note: 'Critics who explain everything by contemporaneity are as bad as those who explain everything by influences. In either case there is a glut of evidence, but is it evidence?'

at first sight seems formless and prosy, we strike patches almost too deliberately lifted from his private classics. Thus:

> These with a thousand small deliberations
> Protract the profit, of their chilled delirium,
> Excite the membrane, when the sense has cooled,
> With pungent sauces, multiply variety
> In a wilderness of mirrors.[55]

This is very neatly on the model of Mr Eliot's favourite passage from Tourneur.[56] Eliot in his earlier poems was addicted to a monumental kind of wit; later his style has been more diffuse, sometimes reminiscent of Pound in *Cathay*:[57]

> Then at dawn we came down to a temperate valley,
> Wet, below the snow line, smelling of vegetation;
> With a running stream and a water-mill beating the darkness,
> And three trees on the low sky,
> And an old white horse galloped away in the meadow.[58]

In his latest verse, the choruses in *The Rock* (1934), the Bible has (appropriately) been an important stylistic factor. Eliot is as receptive to stylistic influences as Joyce. Thanks to a stern self-discipline he has avoided writing mere pastiches but a less wary poet who tries to adopt the Eliot way of writing will find it is not one way but many and be lost in derivativeness. What first strikes readers of Eliot is his gift for point and his surprise effects; but point and surprise can both be faked. In *The Rock* Eliot himself seems to me to get some of his smart effects, especially of antithesis, too easily; they are too pat; they do not ring true. This is an inevitable trap for his imitators.

Eliot's verse was only free in that he allowed himself to ring the changes quickly; one moment conversation; the next moment Senecan sententiousness (for the latter the traditional blank verse

[55] 'Gerontion' (1919), ll. 62–6, with allusion to lascivious mirrors from Ben Jonson's *The Alchemist* (1612): Grover Smith, op. cit. n. 25 above, 64.

[56] Cyril Tourneur, *The Revenger's Tragedy* (1607–8), III. v. 69–79, quoted by Eliot in 'Philip Massinger' (1920), *Selected Essays*, 209, and cited by Spender, *Destructive Element*, 141, in reference to another passage in 'Gerontion' beginning 'After such knowledge, what forgiveness?' (ll. 33–6). Cf. *Modern Poetry*, 145.

[57] Ezra Pound's verse adaptations of Chinese poems by Li-Po (Rihaku) were collected in *Cathay* (1915), later in *The Translations* (1953); these were outstanding examples of free verse.

[58] 'Journey of the Magi' (1927), ll. 21–5.

line is always cropping up again; but thanks to quick change juxtapositions it had temporarily regained its freshness). But other poets were bound to want a more stable medium. There were two ways of obtaining this; to loosen the verse more (as in *Cathay*, 'The Journey of the Magi', etc., Walt Whitman and Blake's Prophetic Books) or to tighten it up to the tension of blank verse or heroic couplets by devices which would make it appear new. Abstractly these are alternatives; actually the New Poets have compromised between them. An example of loosening without tightening is Mr Sacheverell Sitwell who uses a resolved verse somewhat like Blake's and sustains it through long poems without it falling to pieces but at the same time without achieving any powerful effect. At its best it is simple and adequate:

This mill that was true harvester now lifts his flail in vain for them,
And the stream slaves for ever and can have no rest.
The cockerel and his fanfare to their last march of sun
Stood on that stone bridge, the highest station for his watch,
And was moved in a hen-coop before the sun rose high.[59]

Here again we cannot talk of technique without bringing in content; Mr Sitwell enumerates too much; he has no economy. Economy is the key-word to the best modern poetry and it is that which resolves the loosening and tightening processes into one. I give three examples of simple statement; notice the reappearing beat of the old verse, which yet, thanks to careful restraint, does not override their essential novelty; notice also what the three have in common.

(1) From Clere Parsons:

> Memory obfuscates and fancy obscures
> also these sentences which slacken and pause;
> causeless intrude the Quarter Boys of Rye
> and early golfers walking on Camber dunes.[60]

[59] 'The Eckington Woods', ll. 138–9, 141–3: *Collected Poems* (1936), 350.

[60] 'Interruption', ll. 17–20: *Poems* (1932), 15. Clere Parsons (1908–31) was, with Auden and Spender, one of the most prominent undergraduate poets at Oxford in 1928: MacNeice, *The Strings are False* (1965), 113–14. Cf. *Modern Poetry*, 149. See also Geoffrey Grigson, *Recollections Mainly of Writers & Artists* (1984), chap. 5.

(2) From W. H. Auden:

> You talk to your admirers every day
> By silted harbours, derelict works,
> In strangled orchards, and the silent comb
> Where dogs have worried or a bird was shot.[61]

(3) From Stephen Spender:

> Then those streets the rich built and their easy love
> Fade like old cloths, and it is death stalks through life
> Grinning white through all faces
> Clean and equal like the shine from snow.[62]

In these examples the absence of 'cleverness' is notable, and the loosening is more evident than the tightening. Before enumerating the tricks which belong to the tightening I will remark on the affinities of this new verse with modern prose.

Prose nowadays has lost its graces; similes and metaphors are comparatively rare; it shuns the purple patch and the personal digression. Writers who keep up an appearance of *Belles Lettres*, whether playful or serious appear unreal. In reaction from æstheticism we write our everyday prose to very simple recipes; we feel that in Walter Pater style and ideas were not fused together; his prose is indigestible.[63] We avoid conscious pattern. But pattern will always assert itself; our omissions and the abruptness of our syntax acquire positive value; e.g., the following: 'I found a bookworm, no longer young, living from home, a mainlander, city-bred and domestic. Married but not exclusively, a dog-lover, often hungry and thirsty, dark-haired. . . . No farmer, he had learned the points of a good olive tree. He is all adrift when it comes to fighting, and had not seen deaths in battle. He had sailed upon and watched the sea with a palpitant concern, seafaring being not his trade. As a minor sportsman he had seen wild boars at bay and heard tall yarns of lions. . . . ' (From T. E. Shaw's introduction to his translation of the *Odyssey*).[64] One need only

[61] 'Consider this and in our time' (1930), ll. 21–4: *The English Auden*, 47.

[62] 'After they have tired of the brilliance of cities', ll. 5–8: *Collected Poems 1928–1953* (1955), 34.

[63] Since T. S. Eliot's precise prose, Walter Pater's (1839–94) elaborate late-Victorian prose has become precious and stale.

[64] T. E. Lawrence (pseud. 'T. E. Shaw') referring to Homer in 'Translator's Note' to *The Odyssey* in his trans. (1932), ii.

imagine how the essayist of the old school would have spread this out and smothered it with butter. The elegant austerity of this prose (relying largely on asyndeton) is well matched in the verse of W. H. Auden. But verse, unlike prose, cannot attain adequacy merely by avoiding padding; the poet's object is not merely to tell you certain facts. Verse therefore will need more tricks than prose.

The most common tricks in Auden and his fellows (who however different from him, have gone to him for a model) are the following (which I mention in a random order):

(a) Counterpointed Rhythm (with occasional imitation (1) of G. M. Hopkins's so-called 'sprung' rhythm,[65] (2) of Skelton's rhythms as revived by Robert Graves in his *Marmosite's Miscellany*).[66] Need I say here what I have already implied, that rhythm in poetry is a different thing from rhythm in music? In poetry you cannot eliminate the pull and thrust of *meaning* ; e.g. 'Put up your bright swords or the dew will rust them.'[67] If the poet ignores this fact, his versification will be coarse (as much of Swinburne).

(b) Assonance in the place of rhyme as used by Wilfred Owen, e.g.,

> But cursed are dullards whom no cannon stuns,
> That they should be as stones.[68]

(c) Merely vowel assonance: (blood, sun).

(d) False rhymes as used by Yeats, Owen, and many others: (blood, cloud; drop, up).

(e) Rhymes of the kind much used by Robert Graves: (may, caúseway; stick, prophétic).

[65] Common (running) rhythm, some of it counterpointed, is distinguished from sprung rhythm by G. M. Hopkins in 'Author's Preface' to *The Poems*, 4th edn corr. (1970), 45–9. Cf. *Modern Poetry*, 122.

[66] Robert Graves (pseud. 'John Doyle'), *The Marmosite's Miscellany* (1925), exploiting the Skeltonics or 'tumbling verse' of the Tudor poet John Skelton (1460–1529)—brief, headlong lines related to doggerel.

[67] Shakespeare, *Othello* I. ii. 59: 'Keep up your bright swords, for the dew will rust them'.

[68] 'Insensibility', 1917–18, st. 6: *The Complete Poems and Fragments*, ed. Jon Stallworthy, 2 vols. (1983), i. 146.

(f) Feminine false rhymes as used in Pound's *Mauberley*. (Τϱοίῃ, lee-way).[69]
(g) Internal rhymes. (Day-Lewis's speciality).
(h) Alliteration (often dominant, on the Anglo-Saxon model).
(i) Pauses (mental and metrical), cp. 'Free Verse'.
(j) Broken or accelerated syntax (cp. Joyce).
(k) Telegraphic omission of minor parts of speech such as articles and pronouns. (Auden's speciality).
(l) Omitted punctuation (for the sake of continuity).
(m) Omitted capitals at the beginning of lines (ditto).
(n) The above are *technical* innovations in the narrow sense, but these poets also innovate in the technique of content, e.g., instead of metaphor or simile imposed *ab extra* they use what is rather the concrete instance of an immanent theme. This can be seen best in Auden.

The above tricks, which I have mentioned at random, had previously been exploited for their own sakes by experimenters; the New Poets, as a rule, take them in their stride, confidently. It can be maintained that they still overdo some of the tricks, e.g., false rhyme; that false rhyme has a certain psychological effect (a kind of heroic bathos, as found in Yeats) and that these poets use it indiscriminately and therefore viciously. Here I would refer to what I said about expression and fashion. It might have been said of the eighteenth-century heroic couplet that it was only adapted to certain effects or only capable of expressing a very limited sphere, that therefore the eighteenth-century poets used it indiscriminately and viciously (as was maintained by the nineteenth-century Romantics). But here, at the risk of appearing a fatalist, I would reply that the eighteenth-century poets could not help using the heroic couplet and that, if one must talk of 'expression', the heroic couplet was the best expression of their world because it was a powerful fashion and powerful fashions are *constitutive*.

Passing from their technique, which we have isolated for our purposes, let us isolate also the content of these New Poets. We could not do this if they were exponents either of Pure Music

[69] 'Hugh Selwyn Mauberley (Life and Contacts)' (1919–20), sect. 1, st. 1, offers a trick rhyme between Greek ('Troy') and English: *Personae: Collected Shorter Poems* (1952), 197.

or of any kind of automatism.[70] For in Pure Music the content should be nothing, and in automatism it might be anything. But these poets are reactionary; they not only have themes but they have a partiality for a special kind of theme. Just as they are acquiring a new poetic diction or poetic syntax which is used without questioning (a good thing; poets have not eternity to make up their minds in) so they are acquiring a new poetic subject matter which, like classical mythology, will be amenable to a variety of poets and comprehensible by a variety of readers. The New Poets have a theme and treat it, as a rule, directly. This is another reaction; allusion and suggestion are still used but they are subsidiary; the New Poets tend to mistrust a mystical obscurantism (e.g., the Abbé Brémond or Mr A. E. Housman's recent lecture on *The Name and Nature of Poetry*).[71] This is where their propagandist element has helped them; you cannot propagand on a *Symboliste* basis. Those who say that prose describes while poetry suggests, are making too easy a distinction. You must walk before you can dance; you can't be a master of suggestion unless you are a master of description. Description is always present (implicitly) in suggestion, as walking is in dancing. To describe anything well is difficult and the New Poets realize, as some of their predecessors did not, the necessity of hard work.

Three notable elements in the poetry of Auden, Spender, Day-Lewis, etc., are the topical, the gnomic, and the heroic. These elements qualify and reinforce one another. To be merely topical, by mentioning carburettors or complexes, is not truly modern. To be merely gnomic, without a concrete foreground, is boring and ineffective. To be merely heroic is escapism, shallow decoration. The value of a poem consists in a ratio. The ratio is, naturally, difficult to maintain. Day-Lewis e.g. is sometimes too facile in introducing his culverts, gasometers, cylinders and other modern properties; similarly Auden in enumerating (often for purposes of satire) his heterogeneous strings of things or

[70] Poetry aspiring 'towards the condition of music' (Walter Pater's essay on Giorgione, 1873) is associated with the notion of 'pure poetry' advocated by Paul Valéry, George Moore, Abbé Henri Brémond in the 1920s, while automatic writing became fashionable among dadaist and surrealist writers of the same period.

[71] Henri Brémond, *Prière et poésie* (1926), trans. A. Thorold, *Prayer and Poetry* (1927), and A. E. Housman, *The Name and Nature of Poetry* (1933), both espoused a proto- or quasi-mystical notion of poetry.

people who are 'news'.[72] In these three poets there is a strong personal element (contrast the professions of Eliot); thus *The Orators* by Auden, an admittedly personal work, is made very obscure by a plethora of private jokes and domestic allusions.[73] The personal element is a bridge between the topical and the heroic; these poets make myths of themselves and of each other (a practice which often leads to absurdity, e.g., Day-Lewis's mythopœic hero worship of Auden in *The Magnetic Mountain*).[74] This personal obsession can be collated with their joint communist outlook via the concept of comradeship (see again Mr Roberts's preface to *New Signatures*). Comradeship is the communist substitute for bourgeois romance; in its extreme form (cp. also fascism and youth-cults in general) it leads to an idealization of homosexuality. Here we are reminded of Mr Roberts's remark about the 'Greek' quality of these poets.[75] They are in several aspects more 'Greek' than any English poets for some time, e.g. in this concern of theirs with comradeship, in their parade *à la* Greek chorus of heroic or fatalistic truisms (see Auden and Day-Lewis *passim*), in their careful pruning (excepting perhaps Spender) of exclamatory sentimentality. They show, however, an un-Greek vindictiveness which they inherit from D. H. Lawrence. Having so far spoken of the common qualities of these poets as a group, I will try to indicate the importance of Auden and Spender as individuals.

By far the most important work yet produced by any of this group is Auden's *Charade, Paid on Both Sides* which appeared first in *The Criterion* in 1928.[76] It can be very profitably contrasted

[72] MacNeice's note: 'Auden is a journalist poet (I do not mean journalistic). If the *Odyssey* is the work of a longshore Greek [see the reference to Homer by 'T. E. Shaw' above, with n. 64], and *The Winding Stair* is the work of a crank philosopher [W. B. Yeats], Auden's poems are the work of a journalist.'

[73] *The Orators: An English Study* (1932) is 'damaged by notorious obscurity': Monroe K. Spears, *The Poetry of W. H. Auden* (1963), 45, esp. in the private 'Journal of an Airman' (Book II) with its self-indulgent complexities. See *The English Auden*, 59–110.

[74] Day-Lewis's *The Magnetic Mountain*, dedicated to Auden, is generally addressed to Auden, esp. iii, poem 16: 'Look west, Wystan, lone flyer, birdman, my bully boy!'

[75] 'The Greek conception of good citizenship': Michael Roberts, 'Preface', op. cit. n. 42 above, 19.

[76] *Paid on Both Sides*, written 1928, was submitted at end of Dec. 1928 to T. S. Eliot and published after a year's delay in Eliot's *Criterion* 9: 35 (Jan. 1930), 268–90: app. 1, *The English Auden*, 409.

with *The Waste Land*. It is tragic where *The Waste Land* is defeatist, and realist where *The Waste Land* is literary. *The Waste Land* cancels out and ends in Nirvana; the *Charade* (cp. once more the Greeks) leaves you with reality, an agon (see the final chorus).[77] The obscurity of the *Charade* is not the obscurity of *The Waste Land* but mainly an obscurity of method (condensed syntax, etc.). Auden has been very much influenced by the Icelandic Sagas; the plot of *Paid on Both Sides* is the simple saga plot of the vendetta; a chorus does the moralizing. The result is an illusion of history. Poetry to-day is seen (contrast Aristotle with twentieth-century philosophers) to have affinities with history.[78] History is not, for most people, a science; they read it because they take its persons and events as symbols. But symbols of what? The whole point, perhaps, is that we do not know what they symbolize. Philosophies of history over-simplify and cheapen, just as psycho-analysis tends to over-simplify and cheapen our dreams. In his *Charade* Auden is not (not obviously at any rate) grinding an axe; it consequently has a higher kind of value than his satirical poems. (True to type; for tragedy is specifically higher than satire.)[79] We are still suffering from Shavianism, the heresy that the highest work of art is the pamphlet;[80] it will be a great pity if Auden declines into this. The desire to show off your opinions, like all forms of egoism, is useful as yeast; it must not

[77] Contrast the *shantih* or 'peace' at the end of *The Waste Land*, l. 433, with the violent *agon* ('contest') at the end of *Paid on Both Sides*, final chorus, 'Though he believe it, no man is strong': *The English Auden*, 17.

[78] As a result of the later nineteenth-century æsthetic movement culminating in Symbolism, poetry had become associated with intuitive philosophy rather than factual history in the writings of such typical twentieth-century idealist philosophers as Benedetto Croce and George Santayana, *contra* Aristotle, *Poetics*, 9, 1451^{b} 27–32: 'The poet should be a maker of his plots more than of his verses, insofar as he is a poet by virtue of his imitations and what he imitates is actions. Hence even if it happens that he puts something that has actually taken place [i.e. an historical subject] into poetry, he is none the less a poet; for there is nothing to prevent some of the things that have happened from being the kind of things that can happen, and that is the sense in which he is their maker': trans. Gerald F. Else (1967), 34.

[79] Aristotle ranks satire (lampoon) as an inferior kind of art because it represents the particular rather than the general (universal): *Poetics*, 9, 1451^{b} 12–15: op. cit., 33.

[80] G. Bernard Shaw's pamphleteering was notorious; recent examples were *The Adventures of the Black Girl in Search for God* (1932) and *The Political Madhouse in America* (1933).

be your main ingredient. Auden, the journalist, runs the danger of merely showing off, of pamphleteering. I suspect that some of the surrealists are in the same position; if they could remain passive they might be artists but they are too deliberately wire-pulling their reflexes; they have made of their unconscious a kind of political platform.[81] If you want to give your unconscious a chance you must keep your eye on something else. Your unconscious and your opinions are both important, but neither is the main concern of your poetry; poetry lies between, though not cut off from either. Poetry consists in a ratio.

Auden's great asset is curiosity. Unlike Eliot, he is not (as a poet) tired. It is significant that in *The Rock* where Eliot attempts to give his poetry a social reference, the passages which ring truest are still those which are non-social, individualist, even suicidal. (There is nothing so individualist as suicide.) Auden is not so sophisticated; he is not old with reading the Fathers. He reads the newspapers and samples ordnance maps. He has gusto, not literary gusto like Ezra Pound, but the gusto which comes from an unaffected (almost ingenuous) interest in people, politics, careers, science, psychology, landscape and mere sensations. He has a sense of humour. To say he is an Aeschylus as some people have done, is merely stupid and might encourage him to be pompous. His job is to go on observing things from his very unusual angle and recording them (need I say that the combined process of observing and recording = creation?) in his very individual manner. His style is still changing, towards a wider intelligibility; and he has a strong tendency towards satire and burlesque. He may therefore be expected to produce either further 'serious' work as in the *Charade* and the earlier poems, or comic work ranging from mere satire to Aristophanic fantasy. His recently published little play, *The Dance of Death*, was not interesting as verse, but it is a good sign that his eye was on the stage;[82] the new poetic drama, if it is to exist, must have

[81] Probably referring to Charles Madge's poetry and criticism in *New Verse* 1933–5, esp. his article 'The Meaning of Surrealism', *New Verse*, no. 10 (Aug. 1934), 14: 'In placing itself at the service of the revolution, surrealism can be allowed no political reservations.'

[82] Auden's *The Dance of Death* (1933, produced Feb. 1934) was written for the Group Theatre which he had founded in Feb. 1932 with Rupert Doone (pseud. of Ernest Reginald Woodfield, 1903–66) and others: Michael J. Sidnell, *Dances of Death*

entertainment value. Auden's versatility and fertility are invaluable at a time when too many writers are hampered by the fear of not being modish.

Stephen Spender is very different from Auden, who has, however, greatly influenced both his outlook and manner of writing. He has been advertised as the 'lyricist' of the new movement and is the most likely of these poets to become 'popular'. He will also be the man for posterity, if our poetry is ever dug up in fragments; here is someone who really *felt*, posterity will say; and will conclude that he died young. His poems have a fragmentary appearance as it is. I sometimes think that this is vicious but prefer to conclude that it is their virtue. His poems have not got that crystal self-contained perfection which is so glibly attributed to the ideal lyric. Nor do they impress one with the approved shock at a first reading. Their machinery is creakingly evident; the last line of a poem tends to be an especially telling one, while the personal or propagandist (and in either case not very unusual) subject matter is enlivened with 'poetical' images (roses, stars and suns) or with save-work epithets like 'beautiful' and 'lovely'. As for the cultural background, Spender has swallowed D. H. Lawrence whole and mixed him up with Shelley, Nakt-Kultur and Communist Evangelism.[83] Yet, if you read Spender's one volume of *Poems* (published 1933)[84] through several times, you will probably decide that he is an interesting and valuable poet. An American critic, claiming that Spender is over-rated, calls for 'a re-inspection of his performance, a re-inspection of his thoroughly conventional, "poetical" idiom, his relaxed rhythms, and his thin, almost feminine, subject matter.'[85] This is a point of view which it is only too easy to take up. He has not Auden's fecundity or felicity; he is patently very limited. But this being so, he has made capital out of his limitations. The poet's 'hands' are in this case not deft and virtuoso but they have a patient tact (like that of Cézanne) which presses his confused world of

(1984), chaps. 2–3. MacNeice's trans. of Aeschylus' *Agamemnon* (1936) and his play *Out of the Picture* (1937) were also performed by the Group Theatre in Nov. 1936 and Dec. 1938: Sidnell, chap. 10.

[83] See Spender, *The Destructive Element*, chap. 9 ('Notes on D. H. Lawrence').

[84] Spender's *Poems* (1933) was expanded in a 2nd edn (1934).

[85] Robert Penn Warren, 'Twelve Poets', *The American Review* 3 (April 1934), poetry suppl., ed. Allen Tate, 227.

emotional clichés into a harmony which is, fittingly, incomplete. His 'relaxed rhythms' are entirely suited to his subject matter; slickness is alien to him. I find myself that when I want to make sure not to be fulsome, I tend to write in this way. Thus two lines of mine which I consider good—

> These moments let him retain like limbs,
> His time not crippled by flaws of faith or memory[86]

are, if I am not mistaken, typical of Spender's manner. It is inevitable, as with most 'emotional' poets, that some of his poems should be failures, should appear merely bald and therefore blatant statements of personal feeling (economy has the same pitfalls as bravado). Spender is an individual poet; individuality (cp. again Cézanne) is always in the making; Spender works hard at his job; that is why his poetry at first sight is incomplete but at second sight necessarily so. It is unconvincing to analyse him. All the features of a face may be 'bad' but the face itself 'beautiful'.

There are a dozen or so other young poets in England who, whatever their 'creed', at any rate believe in, and write about, something other than their own moods. I have not spoken of them because, having only seen their work so far in periodicals, I am more struck by what they have in common than by their individualities. There seems also to be a flourishing school in America, which, in reaction from experimental impressionism, is consciously limiting itself to a certain well-girdered and substantial way of writing.[87] Again, I have not read them enough or re-read them enough, to criticize them. For me the history of post-War poetry in England is the history of Eliot and the reaction from Eliot. By Eliot I mean the Eliot of *Poems 1909–1925*. Needless to say Eliot himself has been reacting from his earlier poetic self in *Ash Wednesday*, his various Ariel poems, and, most recently, in the choruses in *The Rock*. But though these later poems may as *kinds of poem* have more possibilities for the future, I feel at

[86] 'Ode' (1934) on the birth of his son Dan in May, st. 21: *The Collected Poems*, ed. E. R. Dodds (1966, corr. 1979), 58.

[87] The 'Fugitive School' of American Southern poets, including John Crowe Ransom, Allen Tate, Donald Davidson, and Robert Penn Warren (see n. 85 above), were anti-romantic poet-critics influenced by T. S. Eliot and I. A. Richards, dedicated to traditional verse forms, who established the movement of 'New Criticism' in new journals like the *Southern Review* (1935–42; 1965–) and later the *Kenyon Review* 1938–68; 1979–) and the *Sewanee Review* (1944–).

the moment that 'Gerontion', *The Waste Land*, *The Hollow Men*, and even 'Prufrock' are better *of their kind*. Eliot is a theologian *manqué* ; he is very susceptible to myth; his myths make him as much as he makes them. *The Waste Land* is a very delicate dovetailing of his favourite myths, profane and sacred. He has written much more simply and clearly from the public's standpoint but not perhaps from his own. On this question of obscurity, I would say finally that, though it is wrong to go out of one's way to be obscure, it is just as wrong to go out of one's way to be intelligible. The poet must fulfil (a far better word than 'express') himself. Eliot fulfilled himself in *The Waste Land*. He is essentially that kind of poet. *The Waste Land*, like *The Ancient Mariner*, cannot become a practical classic, i.e. a classic which the next poets shall use as a model. I have written this article primarily to indicate how the new English poetry is developing. This is why I have spoken of Eliot's poetry mainly as *a point from which* instead of assessing it on its own undeniable merits. This is also why I have not spoken of our other great living poet, Mr Yeats.

Mr Yeats is the best example of how a poet ought to develop if he goes on writing till he is old. I am not one of those who have nothing to say for his earlier poems and everything to say for his later poems. He is a very fine case of identity in difference and anyone, who is not pleading some irrelevant cause, can see this; e.g., to mention details, he still speaks of 'Tully' and he still uses the fantastic refrain *à la* D. G. Rossetti or Morris.[88] But he has, in his own way, kept up with the times. Technically he offers many parallels to the youngest English poets. Spender is like him in that they both have worked hard to attain the significant statement, avoiding the obvious rhythm and the easy blurb. Auden and Day-Lewis both use epithets in Yeats's latest manner. But when all is said, Yeats is esoteric. He is further away from the ordinary English reader or writer than Eliot is; not only because of his cabalistic symbols, etc., but even more because of the dominance in him of the local factor. His rhythms and the texture of his lines are inextricably implicated with his peculiar past and even with the

[88] Yeats referred to Marcus Tullius Cicero as 'Tully' in 'Mad as the Mist and Snow' (*Words for Music Perhaps*, poem 18) and used 'the fantastic refrain' esp. in his Crazy Jane poems and political lyrics of the 1930s. Cf. D. G. Rossetti's (1828–82) 'Sister Helen' and William Morris's (1834–96) 'The Wind' and 'Two Red Roses across the Moon' for use of ballad refrain.

Irish landscape. They are, therefore, not to be too closely copied. If we must copy we should either copy people of our own age and society (wholesome plagiarism) or else people so far removed from us by time or language that our copying will not impose upon anyone. Thus I avoided reading Eliot for three years or so, that I might not write fake Eliot. But when I wrote the Spenderesque lines quoted above, I did not regard them as fake Spender because I knew that they were proper to me; whereas Eliotesque lines would not have been proper to me but merely composed *ab extra*.

I will end with a few predictions for the near future. 'Pure poetry', as we have seen, is on the decline; this does not mean that poetry is to masquerade as anything else. For poetry *qua* poetry is an end and not a means; its relations to 'life' are impossible to define; even when it is professedly 'didactic' 'propagandist' or 'satirical' the external purport is, ultimately, only a conventional property, a kind of perspective which many poets like to think of as essential. In the near future we shall have a great deal more communist poetry (and a certain amount of reactionary nationalist—witness Mr Hugh MacDiarmid).[89] But in as far as these poets have the sense of touch, their poetry (whether they think so or not) will not be ancillary to their politics, but their politics ancillary to their poetry. The same applies to psychology. Poets may believe that a poem is merely a psychological document, and such a belief may encourage their output but, in as far as they have the sense of touch, their psychology will be subordinated. I see a future, therefore, for both psychological and political poetry, though in both classes there will inevitably be a good deal of faking and a good deal of honest futility.

What I have said about 'History' should indicate the importance of narrative poetry. I have noticed that most theories of what poetry is, leave no room for Homer, though nearly everyone assumes that Homer is poetry. This inconsistency is due to the dominance in modern times of the lyric; disgracefully fostered by both dilettantes and mystagogues. The short poem has naturally a unique concentration; even so I notice that, in spite

[89] Christopher M. Grieve (pseud. 'Hugh MacDiarmid', 1892–1978), *First* and *Second Hymns to Lenin* (1931, 1935).

of the popular assumption that any one 'lyric' should be entirely self-supporting, yet our appreciation of it is greatly helped by a reading of other lyrics by the same author; more than this, we think of it (generally half-consciously) as a part of a whole, constituted by its author's other poetry and often (if we are not to cant) by certain aspects of the author himself.

Spender's latest poem, *Vienna*,[90] is a poem of nearly thirty pages on a contemporary theme. Unsuccessful, I think (still too *voulu*) but the right kind of experiment. I expect poets in the near future to write longer works (epics, epyllia, verse, essays and autobiographies) which will, for the intelligent reader, supersede the stale and plethoric novel. The more sensitive novelists (e.g. Mrs Virginia Woolf) are approaching poetic form.[91] The novel-reading public will always gulp its port, and writers who want their historicized symbols to be taken seriously would be better advised (provided they have 'a feeling for words') to deploy them in verse.[92]

As for Poetic Drama, there are at the moment many signs of a renaissance.[93] Here, as in narrative, the verse-technique will at first have to be rather loose, until a compromise is reached between significant condensation and more or less instantaneous intelligibility. In its *dramatic* aspect it will probably go through an eclectic stage (borrowing from the music hall, morality plays, etc.) until, in Aristotle's phrase, it attains its own nature.[94]

In narrative, drama, propaganda, satire, etc., we shall have to compromise with the public. In the shorter poem not so much so. It would be a great pity to go back to the Georgians and

[90] (1934).

[91] After James Joyce's *Ulysses* (1922), Virginia Woolf's *To the Lighthouse* (1927) and *The Waves* (1930) were noteworthy for 'poetic form'.

[92] Three years later MacNeice followed his own advice, to deploy 'historicized symbols' in novelistic verse, by writing *Autumn Journal* 1938 (1939).

[93] In addition to Yeats's recently published *Collected Plays* (1934) and *A Full Moon in March* (1935) and to Eliot's recent *The Rock* (1934) and *Murder in the Cathedral* (1935), there was in 1935–6 a season of Group Theatre presentations at the Westminster Theatre, London, opening on 1 Oct. 1935 with Auden's *The Dance of Death* and Eliot's *Sweeney Agonistes* (1932): Sidnell (above, n. 82), chap. 6.

[94] τὴν αὑτῆς φύσιν ('its self-nature' or natural form): Aristotle, *Poetics*, 4, 1449^{a} 13–5: '[Tragedy] did expand gradually, each feature being further developed as it appeared; and after it had gone through a number of phases it stopped upon attaining *its full natural growth*': trans. Else, 22 (editorial italics indicate MacNeice's allusion). Cf. 'Letter to Graham and Anna', 16 Aug. 1936, l. 95: *The Collected Poems*, 63.

sacrifice the intellectual gains of the 'difficult' poets. But I imagine we shall allow our readers some compensation. When we are esoteric (mystical or metaphysical) we shall prop them up with a palpable outward form (like Yeats or Valéry) till they have time to collect our tenuous implications. When, on the other hand, we are chaotic in outward form we shall give them something to grip in the nature of 'content'. The best poems are written on two or more planes at once, just as they are written from a multitude of motives. Poetry is essentially ambiguous, but ambiguity is not necessarily obscure. There is material for poetry everywhere; the poet's business is not to find it but to limit it. Part of his job is forgetting. We want to have the discoveries of other poets in our blood but not necessarily in our minds. We want just enough a priori to make us ruthless so that when we meet the inrush of a posteriori (commonly called 'life') we can sweep away the vastly greater part of it and let the rest body out our potential pattern; by the time this is done, it will be not only a new but the first pattern of its kind and not particularly ours; the paradox of the individual and the impersonal. To write poetry needs industry and honesty and a good deal of luck.

BOOKS TO READ

Besides the poems of T. S. Eliot, Ezra Pound, W. B. Yeats (particularly those in *The Tower* and *The Winding Stair*), Robert Graves, Laura Riding, Edith and Sacheverell Sitwell, D. H. Lawrence, Roy Campbell (a free lance who makes something of high falutin),[95] W. H. Auden, Stephen Spender and Cecil Day-Lewis, it is important among dead poets to read Gerard Manley Hopkins and Wilfred Owen. Hugh MacDiarmid's *A Drunk Man Looks at the Thistle* is a vigorous fantasia in synthetic Scots.[96] Wyndham Lewis's *One-Way Song*, a long didactic poem in vigorous rhyming couplets, should be read as the only one of its kind.

The chief anthologies are *New Signatures* and *New Country*,[97] both edited by Michael Roberts—the latter contains work by Charles Madge,

[95] Roy Campbell (1901–57), the South African expatriate poet and satirist, known at this time for his boisterous *Adamastor* (1930), anti-Georgian *The Georgiad* (1931), and several other works.

[96] (1926). [97] (1932, 1933).

R. E. Warner, and Richard Goodman[98]—and *Active Anthology*, edited by Ezra Pound, and instructively bad. It includes poems by William Carlos Williams, Marianne Moore, etc.

Of books on modern English poetry I recommend *A Hope for Poetry* by Cecil Day-Lewis,[99] and *The Destructive Element* by Stephen Spender. Work by most of the younger poets is published in *New Verse* (every two months, 4a Keats Grove, London, N.W.3).[100]

The Newest Yeats[1]

A Full Moon in March. By W. B. Yeats. (Macmillan. 5*s*.)

I do not know how much Yeats was influenced by Synge but his latest verse play, *A Full Moon in March*, reminds me of Synge (astonishing though that may seem) and of what Yeats found in Synge, 'an astringent joy and hardness'.[2] This little play (beside which its earlier version, *The King of the Great Clock Tower*,[3] is merely a decorative piece) is the thin end of the wedge but it has the weight of a life and philosophy behind it and its edge is a very sharp one, a final expression of the necessity of desecration and the bravado of defeat.[4] It must be read several times for at

[98] Rex (Ernest) Warner (1905–86) and Richard (Hacklett) Goodman (1911–) had been Oxford undergraduates, Communist poets, and contributors to *New Country*. Goodman matriculated at New College, Oxford, 1929, did not graduate till 1945, but published *Poems* (1931), co-edited the *Oxford Outlook* 1931 with Sir Isaiah Berlin, and edited *Oxford Poetry 1932*. Warner, a graduate of Wadham College, Oxford, published *Poems* (1937), became a novelist and translator of Greek classics, and was MacNeice's predecessor as Director of the British Institute, Athens, 1945–7. He was a friend of Auden's but never met MacNeice.

[99] (Oxford, 1934; with postscript, 1936.)

[100] *New Verse*, ed. Geoffrey Grigson, 1933–9, to which MacNeice, Auden, and many other poets of the thirties contributed: Grigson, *Recollections*, chap. 6.

[1] *New Verse*, no. 19 (Feb.–March 1936), 16. *A Full Moon in March* first appeared in *Poetry* Chicago (March 1935), then as a book with preface and poems (Nov. 1935).

[2] John Millington Synge (1871–1909), the Irish dramatist, was a great influence on Yeats's work; for the quotation, see Yeats, 'Preface to the First Edition of John M. Synge's *Poems and Translations*' (1909), *The Cutting of an Agate* (1912), collected in *Essays and Introductions* (1961), 306.

[3] Yeats began *The King of the Great Clock Tower* (Dec. 1934) in prose in order to 'be forced to make lyrics' after two barren years: Yeats, *The Variorum Edition of the Plays* (1966), 1310.

[4] At the end of *A Full Moon* the Queen's dance with the Swineherd's severed head is interpreted by an attendant as a descent into 'desecration and the lover's night' (refrain, ll. 185, 191, 197); cf. *The Great Clock Tower* (refrain, ll. 186, 192, 198, 204).

first sight one may say 'There is nothing new here' (true but not entirely). Yeats's study of Swift has born fruit—

> I study hatred with great diligence,
> For that's a passion in my own control.[5]

In 1900 Yeats desired rhythms which should be 'the embodiment of the imagination, *that neither desires nor hates, because it has done with time*'.[6] (Italics mine.) But the metaphysical paradox[7] demands the descent into time; Yeats's later poetry is an open-eyed recognition of that descent.

Of the shorter poems in this book none is as striking as the best in *The Winding Stair* but the Supernatural Songs improve on re-reading.[8] And no one by this stage should have the impertinence to read Yeats only once.

Sir Thomas Malory[1]

Sometimes in dreams the dream becomes palpably more substantial. The process is like scrambling eggs. From an indefinite froth comes, seemingly instantaneously, something with a recognizable

[5] *Supernatural Songs* 5 'Ribh considers Christian Love insufficient' (1934), st. 1. In his last decade, 1929–39, Yeats became obsessed with Jonathan Swift (1667–1745), the great Anglo-Irish satirist, esp. in the one-act play *The Words Upon the Window-Pane* 1930 (1934).

[6] 'The Symbolism of Poetry' (1900), *Ideas of Good and Evil* (1903), in *Essays and Introductions*, 163. A fuller excerpt is quoted in *Modern Poetry*, 22.

[7] The natural-supernatural or object-subject distinction raises the question as to whether metaphysics is indeed possible at all; hence, return from speculative ideas to the 'real world' of objects and experiences is always necessary. Kant was the first philosopher seriously to ask this question in *Kritik der reinen Vernunft* (*Critique of Pure Reason*, 1781, rev. 1787), I, Div. ii, Bk. 2, chap. 2. (Neo-Kantians in 1910–40 formed a strong school of thought—including Ludwig Wittgenstein, C. I. Lewis, Henri Poincaré, R. G. Collingwood.) For MacNeice on Kant, see his review of R. V. Feldman's *The Domain of Selfhood* (Oct. 1934), to be published in the second volume of selected prose.

[8] The twelve Supernatural Songs in *A Full Moon*, expanded from eight in *The Great Clock Tower*, were lyric challenges attacking Western Christian civilization from the viewpoint of Eastern pagan wisdom and written in an allusive 'cosmic' shorthand.

[1] *The English Novelists: A Survey of the Novel by Twenty Contemporary Novelists*, ed. Derek Verschoyle (30 April 1936), 17–28. Among other contributors were H. E. Bates, Elizabeth Bowen, Bonamy Dobrée, Graham Greene, Rose Macaulay,

texture, something one can put in one's mouth. When a dream behaves in this way, it is becoming a work of art.[2] The effect is often one of healthy bathos. From a sickly-sweet twilight of indefinite sensations there emerges perhaps the exceedingly familiar, exceedingly detailed, figure of someone one knows, and this at once makes the dialectic of the dream concrete. So it is with Malory.

Malory has sometimes (rightly but not fairly) been censured for misunderstanding of his originals, prosaic outlook, bathos, poor construction and inconsistent portraiture—for his lack, in fact, of those two supposed essentials, method and a point of view.[3] For this we should be thankful. The novel is not a school-exercise, ten marks for construction, five for characterization, three for the moral, two for the style.

The novel is the furthest removed literary form from the philosophical treatise. Even philosophical treatises can benefit from the random element (is it not perhaps its inconsistencies which have fascinated students of the *Critique of Pure Reason*?)[4] but, generally speaking, a philosophical treatise should be completely under the control of the mind of the author, a novel not so. Of

Edwin Muir, Sean O'Faoláin, V. S. Pritchett, Peter Quennell, L. A. G. Strong. As MacNeice had written a novel *Roundabout Way* (1932, under the pseud. 'Louis Malone'), he fulfilled the editorial requirement of living novelists writing on the novel (Verschoyle, Introduction, xi). For MacNeice, Malory's *Morte Darthur* (1485) had been 'one of my sacred books' from school and college days (*The Strings are False*, 98; see also 77–8, 221). Eugene Vinaver's *Malory* (1929) was his main scholarly source in this article, esp. in the second half.

[2] Cf. a passage in Freud familiar to MacNeice from Auden's quotation of it in 'Psychology and Art To-day', *The Arts To-Day*, 4: 'A true artist . . . understands how to elaborate his day-dreams. . . . Further, he possesses the mysterious ability to mould his particular material until it expresses the ideas of his phantasy faithfully.'—Sigmund Freud, con. to 23rd lecture, 'The Paths of Symptom-Formation', in *Introductory Lectures on Psycho-Analysis*, 1st auth. trans. Joan Rivière, pref. Ernest Jones, 2nd edn (1929), 314–15. Standard edn (1953 ff.) of this passage differs slightly. (MacNeice had contributed the article 'Poetry To-day' to Grigson's *The Arts To-Day*: see above.)

[3] See Vinaver, op. cit. n. 1 above, chaps 3 and 7.

[4] 'Seldom, in the history of literature, has a work been more conscientiously and deliberately thought out, or more hastily thrown together, than the *Critique of Pure Reason*. . . . [Para.] . . . Kant flatly contradicts himself in almost every chapter; and . . . there is hardly a technical term which is not employed by him in a variety of different and conflicting senses. As a writer, he is the least exact of all the great thinkers. [Para.] So obvious are these inconsistencies that every commentator has felt constrained to offer some explanation of their occurrence.' Norman Kemp Smith, *A Commentary to Kant's 'Critique of Pure Reason'*, 2nd edn (1923), xix–xx.

course very often a novelist thinks he is controlling his material when he is not. Even with such a self-conscious artist as James Joyce it looks in *Ulysses* as if it is his *material* which is making the running. Joyce is pre-eminently a selector but, as a true novelist, he selects by touch rather than by theory. Malory of course was a selector in a very narrow sense—namely, a redactor. He took a number of books in a foreign language and made out of them one book very much shorter in his own. If he had been a theorist, one-minded, Shavian, he would have gone about his job quite differently. His book would have come from his hands ready to be labelled—an exposition of artificial romantic love *à la* Chrestien de Troyes,[5] or of medieval Christian mysticism with the stress on the Grael,[6] or of pure chivalry with a fifteenth-century moral, or maybe a Celtic romance weltering in marvels, or a national epic purged of Gallic accretions and with the emphasis on Arthur's insularity, his hardy impudence against the Roman Emperor.[7] But what reached Caxton's press[8] was none of these things.

None because all. The *Morte D'Arthur* is a divine mix-up. If Miss Jessie Weston could have told Malory that the Grael is the superfoetation of Christianity upon a pagan vegetation myth,[9] he would not have understood her but, if he could have understood

[5] Chrestien (Chrétien) de Troyes, the French poet (*fl.* 1170), wrote the earliest courtly romances, which marked the coalescence of the Arthurian story with the Provençal tradition of *amour courtois* (courtly love). For Chrestien, see Vinaver, op. cit. n. 1 above, 17–18, *et passim*.

[6] In medieval legend, the Holy Grail, Saint Graal(e), or Sankgreal was the chalice or dish used by Christ at the Last Supper, treated as a visionary object of quest for knights of the Arthurian romances (Gawain, Perceval, Lancelot, Galahad, Bors) from Chrestien's *Conte de Graal* through Robert de Borron's trilogy and Wolfram von Eschenbach's *Parzival* (in which the Grail is a stone) to Malory's *Morte Darthur*, *c.*1165–1485. (Of course, T. S. Eliot's allusion to the Grail in Notes to *The Waste Land* brought Malory new readers.)

[7] Arthur's expedition against the Roman Emperor Lucius, recounted in Malory's *Morte Darthur*, V.

[8] William Caxton (*c.*1422–91), translator, the first English printer, set up a press in Westminster 1476; he published Malory's book 31 July 1485.

[9] Jessie L. Weston's *From Ritual to Romance* (1920, 1941), source of the Grail story for Eliot's *The Waste Land*, links pagan fertility rituals of vegetation gods of the Middle East as expounded in Sir J. G. Frazer's *The Golden Bough* to Church liturgy via Christ's death and Resurrection as a rite of one of these 'dying gods': see Weston, chaps 4–6 and 12. 'Superfoetation' is a mock-patristic term used by Eliot in his poem 'Mr Eliot's Sunday Morning Service': 'In the beginning was the Word. / Superfetation of τὸ ἕν' [The One]—referring to 'a blasphemous canard that the

her, he would have been very angry. The Grael, with its Cistercian subtleties,[10] was difficult enough to handle as it was. Yet the vegetation myth, whether we call it that or not, is there. The Waste Land and the Wounded King[11] represent something which is vastly old—Jung might call it an archetypal myth[12]—and which the reader appreciates all the better because the symbolism is not explicit. The novelist's job is not to be explicit. To take Joyce once more as an example, Dedalus in *Ulysses* is a failure because it is clear what he is meant to be, Bloom may be meant to be something also but he *is* so incontrovertibly that his meaning is not given a chance to wreck him. Malory's characters are like Bloom rather than Dedalus.[13] Which perhaps explains how it was that Ascham missed the point of the book and said that the whole pleasure of it 'standeth in two speciall points, in open manslaughter and bold bawdrye'.[14] We know what some people think is the whole pleasure of *Ulysses*.[15] It is a virtue in a novelist that his point is able to be missed.

Most people would not classify the *Morte D'Arthur* as a novel, but I cannot see why not. The *Odyssey*, according to T. E. Shaw,

Son, being coeternal with the Father, was superimposed upon His own prior existence in being also begotten'—Grover Smith, op. cit. ('Poetry To-day', above, n. 25), 43.

[10] Cistercians, the White Monks of 12th–13th centuries who cultivated a strict asceticism, were inspired by the mystical theology of St Bernard of Clairvaux. For Cistercian influence on the French *Queste del San Graal*, see Vinaver, op. cit. n. 1 above, 71–8.

[11] In the Grail story, 'the condition of the King is sympathetically reflected on the [waste] land, the loss of virility in the one brings about a suspension of the reproductive processes on the other': Weston (above, n. 9), 21. For the wounded Fisher King, see ibid., chap. 9; for the Waste Land, see ibid., chap. 5, 59–60.

[12] According to C. G. Jung, the great artist (such as the early Christian mystic 'Hermas' or Dante or Goethe) expresses universal human experience out of the 'collective unconscious' in archetypal myths: Jung, 'Psychology and Literature', *Modern Man in Search of a Soul* (1933), trans. W. S. Dell and C. F. Baynes, lecture 8, 152–72. Jung's theory was further expounded by Maud Bodkin, *Archetypal Patterns in Poetry* (1934), a book with wide influence.

[13] James Joyce in *Ulysses* (1922, 1933) contrasts his Telemachus-figure (cerebral Stephen Dedalus) with his Odysseus-figure (earthy Leopold Bloom) in favour of the latter, as Malory in *Morte Darthur* had given much attention to seriously flawed and often earthy characters (Gawain, Tristram, Lancelot).

[14] Roger Ascham, *The Scholemaster* 1570 (Arber edn 1923), 80; (1967) i. 27. Quoted by Vinaver, 114 n.

[15] *Ulysses* had been subject to trial for obscenity in the US District Court, New York, 1933: see Richard Ellmann, *James Joyce*, 678.

was the first European novel.[16] Why not the *Iliad*?[17] Presumably because of that desperate bugbear, 'construction'. I should prefer to call neither of them novels as they are both in verse. But the *Satyricon* of Petronius was a novel, so was the *Golden Ass* of Apuleius and so were the Icelandic Sagas.[18] It is foolishly assumed that the novel, like tragedy in Aristotle's *Poetics*, should possess (a) plot, (b) characterization.[19] But Aristotle's distinction between plot and history[20] will not hold in this genre; the psychologists moreover tell us, what we have always felt in our bones, that there are more kinds of continuity than one.[21] As for characterization no one has ever made clear quite how differentiated or quite how substantiated the characters ought to be in a novel. Malory's claims to be a novelist are so minimal in the eyes of the purists that I will digress to go one worse in their eyes.

I believe that it would be possible to write a novel about two cricketers A and B (those might even be all they had for names) and confining oneself to their matches—scores, strokes and style—yet produce a readable book. Plot here would consist merely in cricket match after cricket match, while the character of A would differ from that of B merely in the external difference of their scores, etc., and the more internal difference of their cricketing technique. Note that in any character, in books or in life, it is impossible to draw a rigid line between his external or incidental attributes—what he does, wears, the shape of his face—and the internal or essential ones, what he thinks, feels. Note again that such a 'plot' as that of my cricketing novel cannot be dismissed as no plot but mere repetition. One cannot merely

[16] T. E. Lawrence (pseud. 'T. E. Shaw'), Translator's Note to *The Odyssey of Homer* (1932), i.

[17] 'Epic [i.e. the *Iliad*] belongs to early man, and this Homer [i.e. of the *Odyssey*] lived too long after the heroic age to feel assured and large': T. E. Lawrence, op. cit., ii.

[18] MacNeice's 'sacred books' at Marlborough College included, besides the *Morte Darthur*, *The Golden Ass* and *Burnt Njal* the Icelandic Saga (*The Strings are False*, 98); Petronius' *Satyricon* became a favourite at Oxford. In 1944–8 MacNeice produced five radio adaptations of these romances.

[19] Plot (*mythos*) and character (*ethos*) are two of six leading elements in Aristotle's theory of tragedy: *Poetics*, 6, 1450^a 3–7.

[20] *Poetics*, 9, 1451^a 38–^{b} 5.

[21] Not only realistic and historical 'plot', but mythic and archetypal 'pattern': see n. 12 above.

repeat. Any enumeration of objects or events will take on a rhythm, as we read it, just as the monotony of the noise of a train takes on a rhythm as we listen to it.

I thought of cricket because I have often found myself taking pleasure day by day in reading the accounts, or merely the batting averages, of players whom I have never seen and do not wish to see. And many people perhaps read everything in the newspapers in this way. Why do they do this? Because, firstly, the newspaper heroes, sportsmen or politicians, are for them dream-figures, *though they know they are real people*; secondly, the repetition with variations of their performances (Mr Ramsay MacDonald . . . Mr Ramsay MacDonald . . .)[22] builds up a vast and gentle rhythm in the back of the mind, hypnotizing us into an escape from reality; but it is the sort of escape people are said to find in opium. They see the real objects and they know they are the same but they see them different. Before any greater claims are put in for Malory, I must repeat my opening point and say that his book throughout gives me the same joy that I get in reading the sporting page in the daily papers or that I get when I see something heroically familiar, however banal, in a dream.

Malory got his heroes out of the French books.[23] They were already a conflation of French and English with a dash of Welsh. Some will say (as Professor Vinaver in his charming book on Malory) that he clipped them into fifteenth-century Englishmen.[24] This is an exaggeration. Malory's characters, more than those of most novelists, are knit up with their background and their background is not stable. The varied threads (how often this book is called a tapestry!) are compromised successfully. Malory is a master of half-conscious compromise, typically English. Thus even his *Quest of the Grael*[25] is acceptable, though we must agree with Vinaver both that the mysticism of it was antipathetic to Malory (after all he was a robber of monasteries), that the Chapels Perilous, etc., were not his natural province, and that the *Weltanschauung* of the Grael, suppressed but not extinguished

[22] J. Ramsay MacDonald (1866–1937), Prime Minister of the first two Labour Governments in Britain, 1924, 1929–31.

[23] Malory refers to his source as a 'Frensshe boke': Vinaver, 29. [24] Ibid., chap. 5.

[25] Malory, *Morte Darthur*, XIII–XVII.

by Malory, clashes very badly with that of the Round Table.[26] But there is much to be said for a clash, even if it is not intended.

Some years ago I used to be most irritated when I came to Book XIII and had to pass from the harpings and joustings of Tristram to the frigid perfection of Galahad. I see now that without Galahad, Lancelot, his unwilling father, would not be so great, or not so *tragically* great, a figure; just as without the Quests preceding it, the Morte proper, the dissolution of the Round Table, would not be a vast moral catastrophe but merely a crash in the cricket averages.

Professor Vinaver is undoubtedly right when he says that Malory wavers continually in his treatment of Lancelot and Guinevere—Malory's 'most cherished ideal is that of happy marriage, and he forgets that marriage and a hero-lover like Lancelot are entirely incompatible'.[27] And Vinaver, quoting the passage where Lancelot condemns paramours, points out that it comes oddly from a man who is 'the very embodiment of adulterous passion'.[28] But is this kind of inconsistency a flaw? It may be that Malory is inconsistent for a purely mechanical reason, that he cannot control his originals. I should prefer to think that he could see two sides of a question, that he could feel with Lancelot in his love as well as in his preaching. But whether he could or not, the fact is that this not too conspicuous vacillating between two or more worlds is something which, whether it is due to skill or accident, is greatly to be desired in novels because it represents, though far more gently, the bitter dialectic of opposites which makes humanity.[29] Any child trained on the Dalcroze system[30] can move two limbs simultaneously in different rhythms. Any civilized man can see simultaneously, or at least in rapid alternation, the point of marriage *á la* Malory and of love *à la* Lancelot.

[26] Vinaver, chap. 6, and App. 3. [27] Ibid., 46. [28] Ibid., 47.

[29] Allusion to G. W. F. Hegel's grandiose dialectical system of opposites in three phases—Logic (thesis), Nature (antithesis), Mind (synthesis), in *Encyklopädie der philosophischen Wissenschaften in Grundrisse* (*Encyclopaedia of the Philosophical Sciences in Outline*, 1817)—*not* to Plato's dialectic of testing an argument through the 'art of conversation'.

[30] Rhythmic education or 'eurhythmics' originated by Swiss teacher-composer Émile Jaques-Dalcroze (1865–1950), founder of schools at Hellerau and Geneva (1910; 1915), influential figure in the development of modern dance.

The main themes and persons of the *Morte D'Arthur* have often been analysed. To illustrate my thesis that Malory to some extent benefited from the difficulties of his task, if not also from the deficiencies of his own mind (it does not hurt a novelist to be something of a zany), I would take the character of Sir Gawaine. Sir Gawaine took my fancy when I first read the book at the age of twelve.[31] I have since tried to find the reason for this. He is not one of the villains, like King Mark or Morgan le Fay, but he more often than not is out of favour with his author.[32] He is a strong knight but not a very knightly one. Perhaps I liked him because I felt that the author had a grudge against him. There are some half-dozen knights who can always beat him, and the reader acquires a sporting hope that he may give one of them a surprise. This sporting interest is a little counterbalanced by the fact that Gawaine has an unfair advantage in that his strength magically increases threefold towards noon. But 'his wind and his evil will' increased with it,[33] which gives him a brute attractiveness. The interest in Gawaine, however, is not merely a sporting one. If Gawaine were absent, not only would the mechanism of the catastrophe (Arthur's war with Lancelot and the revolt of Mordred) have to be contrived anew, but the book would lose a recurring *motif* of moral contrast. Of all the more important knights Gawaine is the furthest from being a paragon.

A curious thing is that in the earlier Arthurian literature he was more or less a paragon, a great national champion—Malory's portrait of him is inconsistent, and how he became as vicious as Malory represents him, I leave to Arthurian scholars.[34] But his viciousness is important. Malory in translating the Queste omitted passages which stressed the opposition of the Grael to the Round Table. He was to the end a hero-worshipper of worldly knights. But Gawaine was one worldly knight, 'a passing hot knight of nature',[35] that Malory makes an example of. His chief characteristic is vindictiveness; the vendetta means more to him than chivalry. He is not merciful, he begins his career by killing

[31] At Sherborne School for Boys, Dorset, MacNeice organized some of his fellows into Knights of the Round Table; he himself took the part of Sir Gawaine: *The Strings are False*, 221.

[32] Cf. Vinaver, op. cit., 73.

[33] *Morte Darthur*, XX. ii.

[34] Vinaver, loc. cit.

[35] *Morte Darthur*, XVIII. iii.

a lady, and he has other bad blots on his record such as the seduction of the lady Ettard.[36] It is notable that it is he who sets the example to the other knights and starts them on the Quest of the Grael, but this he does out of self-glory, not out of holiness, and it only brings disappointment to himself and injury to the Round Table—'Gawaine, Gawaine, ye have set me in great sorrow. For I have great doubt that my true fellowship shall never meet here more again.'[37] Malory concentrates on Gawaine the bitterness which in the original Queste, it seems, was directed against the Round Table in general. Gawaine, in his quest, after a tedious lack of adventure, at last, with two others, is attacked by seven knights, whom he and his companions kill.[38] He did not know that the slain knights were symbols. A hermit, however, reproves him—'ye have used the most untruest life that ever I heard knight live. *For, certes, had ye not been so wicked as ye are, never had the seven brethren been slain by you and your two fellows*' (italics mine).[39] Obviously a new system of values has been slipped into the story. Up till now it had been a good thing to kill knights. We begin to sympathize with Gawaine, who makes no attempt to amend himself and, when the hermit tells him to do penance, refuses—'for we knights adventurous often suffer great love and pain'.[40]

Towards the end of the story Gawaine rises to a kind of brute dignity. He never takes advice and he never forgives. His intractable spirit causes the downfall of Arthur. 'So upon the morn there came Sir Gawaine as brim as any boar, with a great spear in his hand. . . . '[41] He makes Lancelot fight with him against his will, and when Lancelot struck him down, and he could not stand, 'waved and foined at Sir Lancelot as he lay'.[42] Lancelot, of course, was too courteous to strike a wounded man. The contrast here is between the knight-errant *par excellence* and a far more primitive type of hero. Gawaine is like some of the Icelandic heroes.[43] His presence gives the *Morte D'Arthur* a

[36] III. vii; and IV. xxiii. [37] Arthur to Gawaine: XIII. viii.
[38] XIII. xvi. [39] Loc. cit. [40] Loc. cit. [41] XX. xiii.
[42] XX. xxii.
[43] In the summer of this year (1936) MacNeice joined Auden in a tour of Iceland with Bryanston School prefect Michael Yates; see Yates's memoir, 'Iceland 1936' in *W. H. Auden: A Tribute*, ed. Stephen Spender (1974, 1975), 59, 66. MacNeice's 'Eclogue from Iceland' mentions the Sagas (esp. through the ghost of Grettir Asmundson):

necessary taint of earthiness. But it is typical of the beautiful balance of this book that Gawaine is allowed a death-bed repentance—not repentance in general but repentance for his conduct to Lancelot, to whom while dying he writes a practical and dignified letter.[44]

Malory's minor characters are hardly differentiated, but we never question their reality. They are as real as the Wife of Bath though they are not so realistic. Not till Defoe do we get as strong a feeling of reality from English prose fiction. Caxton's preface, by the way, has been taken as evidence that the *Morte D'Arthur* is the first professedly fictitious prose narrative in the language.[45] After Malory's time, of course, many new influences swept away any lingering aversion to non-didactic or unhistoric fiction. But such invented romances as those of Lyly and Sidney[46] show none of the life and solidity of Malory's mere redaction. More important in the history of the English novel were the Elizabethan pamphleteers[47] who anticipated that humour of manners which was later to be so prominent and which, present in Chaucer, was inevitably absent from Malory; Malory shows little humour, though there is a certain folk quality which might be called humour in his story of Beaumains.[48]

In style the Elizabethans produced nothing like Malory's for pure narrative. Lyly's is not, properly, a narrative style at all—'a delicate bayte with a deadly hooke, a sweete Panther with a devouring paunch, a sower poyson in a silver potte'.[49] The pamphleteers made more important innovations, their digressive gusto being the same quality that characterizes so many English novels from Fielding to Thackeray.

Auden and MacNeice, *Letters from Iceland* (1937), chap. 10; MacNeice, *The Collected Poems*, 40–7.

[44] *Morte Darthur*, XXI. ii.

[45] Sir Walter A. Raleigh, *The English Novel*, 2nd edn (1895), 14–15, reiterated by Ernest A. Baker, *The History of the English Novel*, 10 vols. (1924–39), i. 302–3.

[46] John Lyly's *Euphues* (1578) and Sir Philip Sidney's *Arcadia* (1590–8) are cited as romantic fictions of improbable intrigue and elaborate style.

[47] Robert Greene, Thomas Lodge, Thomas Nashe, Thomas Deloney—all 'university wits' who wrote lively prose pamphlets in the 1590s.

[48] *Morte Darthur*, VII, 'The Tale of Sir Gareth of Orkney' (Gawaine's brother, the noble kitchen knave called 'Beaumains'). For Malory's humour, see Vinaver, 65, n. 2.

[49] Euphues' discourse on beauty to Lucilla, in Lyly's *Euphues—The Complete Works* (1902; 1967), ed. R. Warwick Bond, 3 vols, i. 202, ll. 19–20.

The Germaines and lowe Dutch, me thinkes, should bee continually keept moyst with the foggie aire and stinking mistes that arise out of their fennie soyle: but as their Countrey is over-flowen with water, so are their heads alwaies over-flowen with wine, and in their bellies they have standing quagmires and bogs of English beere.[50]

Such writing has robustness and clarity, qualities in which the English novel has never been deficient. Malory's style has something else much rarer, a delicate virility which belongs exclusively to narrative.

This quality cannot be analysed. In some places it seems to be attained by understatement. Malory is not a writer whom we feel *writing* all the time as we hear a clock ticking. Witness the first appearance of the Grael, a passage which Professor Vinaver has censured as inadequate:[51]

so they went into the castle to take their repast. And anon there came in a dove at a window, and in her mouth there seemed a little censer of gold. And therewithal there was such a savour as all the spicery of the world had been there. And forthwithal there was upon the table all manner of meats and drinks that they could think upon. So came in a damsel passing fair and young, and she bare a vessel of gold betwixt her hands, and thereto the King kneeled devoutly, and said his prayers, and so did all that were there. Then said Sir Launcelot, What may this mean? This is, said the King, the richest thing that any man hath living. And when this thing goeth about, the Round Table shall be broken.[52]

This passage is indeed inadequate in that it totally fails to bring out the Grael's significance. As exposition it fails, as mere statement it is magnificent. M. Jean Cocteau says in a note on his play *Orphée*—'Inutile de dire qu'il n'y a pas un seul symbole dans la pièce. Rien que du langage pauvre, du *poème agi*.'[53] It seems to me that Malory attained sometimes this peculiar kind

[50] On drinking, in Thomas Nashe's *Pierce Penilesse his Supplication to the Devill* (1592)—*The Works* (1904–10; 1958), ed. Ronald B. McKerrow, 5 vols, i. 206–7.

[51] Vinaver, 82, in reference to the Fisher King Pelles's apparently worldly estimate of the Grail as 'the richest thing that any man hath living' in the passage quoted by MacNeice immediately below (ll. 9–10).

[52] *Morte Darthur*, XI. ii. The Grail appears first here as a visionary sign just before Lancelot's begetting of Galahad (the destined Grail knight) by Elaine.

[53] See 'Notes on Producing', *Orphée*, trans. Carl Wildman (1933), p. x n.: 'It is not necessary to say that there is no symbol in the play. Nothing but simple language, *acted poetry*.' Quoted again in *Modern Poetry*, 195.

of expression which Cocteau is seeking self-consciously. The philosopher makes a judgement, but the poet and the novelist, on their different planes, make statements. Malory is a master of statement.

Collected Poems, 1909–1935, by T. S. Eliot[1]

Collected Poems, 1909–1935. By T. S. Eliot. (Faber. *7s. 6d.*)

For some time the poetry-reading public has been in need of a new book of collected poems from Mr Eliot, one of the two English poets over thirty-five who still have anything to say to us. It was a serious inconvenience to have to marshal three or four books and the four Ariel poems in order to get a view of Mr Eliot's remarkable poet's progress. The arrangement of these collected poems is chronological and we are enabled to remark simultaneously Mr Eliot's inveterate obsessions and his forceful divergences from his old self. The last poem in this book is a new poem, *Burnt Norton*. Here, after the more direct address of the choruses in *The Rock* and *Murder in the Cathedral*, Mr Eliot, who himself disclaims metaphysical capacity, returns to that kind of metaphysical paradox and antithesis which has always fascinated him. But he returns not as a metaphysician but as a mystic, or—and this is in no way discreditable to him—as a would-be mystic. It is 'the still point of the turning world' that obsesses him—a phrase repeated in *Burnt Norton* from the much-less satisfactory 'Triumphal March'.[2] It is a remarkable thing that the poet who has produced the wittiest social commentary of our period ('Prufrock' and 'Portrait of a Lady') and the most perfect expression of the defeat of our civilization ('Gerontion' and *The Waste Land*) should also produce good poetry written in a spirit of religious acceptance. Mr Eliot is a proof that a dry surface does not necessarily connote a lack of vital sap.

'Shall I at least set my lands in order?' is one of the last lines in *The Waste Land*, and a few lines lower down we have the answer 'These fragments I have shored against my ruins', and that

[1] Unsigned book review in the *Listener* 15: 388 (17 June 1936), 1175.
[2] *Triumphal March* (1931), l. 34; *Burnt Norton* (1935), ii. 16.

is the position in which *The Waste Land* leaves us.[3] It is no wonder that the younger poets, for whom at one time this poem had the authority of a gospel, have after some study of Marx and Freud begun to attempt to order their lands differently. But Mr Eliot is not to be blamed because he has not done so. A man must be loyal to his accidents—the accidents of age, origin and personality—and if some people find it hard to believe that Mr Eliot is sincere in his later reactionary period, in his taking up of Anglo-Catholicism and his vigorous antipathy to 'liberalism', it would be much more difficult to believe in his sincerity if he, 'the man he is',[4] had gone communist along with his juniors.

Mr Eliot has an individual vision and is an excellent craftsman (except that he cannot manage the lighter measures—witness 'Minor Poems'). These two things make good poets. It is doubtful whether anyone will ever 'put across' the Modern City, i.e. modern London, better than Mr Eliot has done. He and Mr Yeats are poets with private outlooks. The private outlook nowadays can only be saved by dignity. No one, it is to be hoped, will deny dignity to either of these isolated, self-critical, and toughly-constitutioned personalities.

Subject in Modern Poetry[1]

The literary critic includes the literary historian, but it is notoriously difficult to write a history of one's own times. Further, a man who writes poetry himself is certain to be prejudiced when

[3] *The Waste Land*, v. 425, 430.

[4] i.e., 'the daily man that he is'—Eliot, *After Strange Gods* (1934), 59, a passage attacking D. H. Lawrence for trusting entirely to 'the Inner Light': 'Of divine illumination, it may be said that probably every man knows when he has it, but that any man is likely to think that he has it when he has it not; and even when he has had it, the daily man that he is may draw the wrong conclusions from the enlightenment which the momentary man has received: no one, in short, can be the sole judge of whence his inspiration comes.'

[1] Essay 8 in *Essays and Studies* by Members of the English Association, collected Helen Darbishire, 22, [Dec.] 1936 (1937), 144–58. Among other contributors were Ernest de Selincourt, Edmund Blunden, C. T. Onions, Dame Helen Gardner.

he comes to criticize poetry. I do not, therefore, profess to attempt a reasoned, balanced, objective, or historical criticism of my contemporaries.

The usual method is to look for, and obviously to discover, literary influences. This method has been over-rated and over-exploited. The fact that I have read A is not what causes me to write α. Just as probably it is because I am the sort of person who is inclined to write α that I choose to read A. Unless I live on a desert island, I could easily shut A at the first page and look for something more congenial.

The literary critic fails through being literary. He is the one kind of writer who really is in the position of Plato's second-hand artist.[2] Literature is a criticism (if in a wider sense of the word) of life;[3] what the so-called critic writes is a criticism of criticism. It is a good sign in the last few years that intellectuals are becoming less interested in books about books. I am told that the prevalence of central heating in American houses makes all Americans catch cold in the open air. So it is with literary literature. Homer, Aeschylus, Bunyan, Dante, did not live in literary self-containedness. Not only the muck and wind of existence should be faced but also the prose of existence, the utilities, the *sine qua nons*, which are never admitted to the world, or rather the salon, of the Pure Artist.[4]

Art for Art's Sake has been some time foundering. A masthead or two even now show above the water with their inconsequent fluttering pennons (the Surrealists?),[5] but on the whole, poets have ceased showing themselves off as mere poets. They have better things to do; they are writing *about* things again.

A similar reaction will probably be noticed in the visual arts. I have recently been to an exhibition of nineteenth-century French

[2] The artist, painter or poet, is an imitator, not a creator, in Plato's *Republic*, Bk. 10, esp. 597E and 598E–599B.

[3] 'Criticism of life': Matthew Arnold, 'The Study of Poetry' (1880), paras. 2–3, in *Essays in Criticism: Second Series* (1888). Cf. article 'Poetry To-day' (Sept. 1935) above, para. 9.

[4] MacNeice's *Modern Poetry* (1938; 1968) was to be a manifesto against pure poetry and pure poetry divorced from life: see his Preface.

[5] MacNeice expands on the subject of surrealism below in this article. Charles Madge and David Gascoyne (1916–) were among few English surrealist poets in the 1930s, although in the 1940s English surrealism became widely known as 'The New Apocalypse': see MacNeice's 'An ABC of Literary Prejudices' (March 1948) below under entry 'Apocalyptics'.

painting[6] and was surprised to find it so unsympathetic. That a harlequin, a plate of apples, and a nude woman bathing should all be presented as equally important, all presented as if with a smirk to say 'Look at me! I can make a work of art out of anything, look at my mastery of form!'—this seems to me a thoroughly unsound attitude. But analogies from the other arts are misleading, and the doctrine of pure form, however vulnerable even there, is far more defensible in painting than in literature.[7] Literature is made with words, and words are a means of conveying a meaning. It is no doubt possible to use words merely for decoration, as the Moors used tags of the Koran to decorate their walls at heights where no one could read them.[8] To do this in literature seems a perversion.

Poetry is a natural or universal activity. We all practise it from our earliest years. Language can be used either scientifically or poetically, and the latter use comes much more naturally to the child and the man in the street. Whenever we say something not merely to record a fact but to record a fact *plus and therefore modified by* our emotional reaction to it (which will involve mannerism in its presentation) we are speaking poetically. Nothing, therefore, could be more vicious than the popular legend that the poet is a species distinct from the ordinary man and that poetry can only flourish in certain places or people, in the highbrow's den or on the slopes of Helvellyn.[9]

[6] An exhibition of 123 19th-century French paintings, organized by the Anglo-French and Travel Society, was held Oct. 1936 at the New Burlington Galleries, London W1: see notice by the *Times*' art critic, 1 Oct. 1936, 10b, with remarks on 'the formal purity of [Georges] Seurat'.

[7] Purism, a protest against Cubist 'destruction of the object', was a modern-art movement led by the French painter Amédée Ozenfant (1886–1966) and Swiss architect 'Le Corbusier' (pseud. of Charles-Edouard Jeanneret, 1887–1965) in their manifesto *Après le cubisme* (1918) and review *L'Esprit nouveau* (1920–5). In *Foundations of Modern Art*, trans. John Rodker (1931, aug. 1952), ii. 300, Ozenfant wrote: 'Pure art is not what is achieved: purity . . . is a maximum efficiency, intensity, and quality, issuing from an utmost economy of means.' Ozenfant moved his School of Fine Arts from Paris to London, 1935–8, then transferred to New York City.

[8] In Easter 1936 MacNeice visited southern Spain with his friend the art critic Anthony Blunt (1907–83) who was 'covering the architecture': *The Strings are False*, 158 ff. Places visited with notable examples of Islamic architecture were Ronda, Toledo, Seville, and probably Granada (not mentioned in *The Strings* passage in which MacNeice was concerned with paintings).

[9] Wordsworth's 'old Helvellyn', mountain in the Lake District southeast of Keswick and a lesser 'Parnassus', is addressed by him notably in *The Prelude*, viii. 1–69.

In its relations to life twentieth-century poetry has had a paradoxical development. It started (the Nineties' tradition diverted into the Georgians) divorced from life but with a fair number of readers. If we do not count the War, when war poems such as those of Sassoon had a large public,[10] poetry subsequently became less popular the nearer it came to life. This was because in its endeavour to be true it became very difficult. Thus Mr Eliot maintained in an essay (1921) that 'poets in our civilization, as it exists at present, must be *difficult*. Our civilization comprehends great variety and complexity, and this variety and complexity, playing upon a refined sensibility, must produce various and complex results. The poet must become more and more comprehensive, more allusive, more indirect, in order to force, to dislocate if necessary, language into his meaning.'[11] But since the time when Eliotesque poetry was predominant, poetry has come yet nearer to life but not in the impressionistic way of the early Eliot. This new poetry is much more intelligible and I shall discuss it presently.

The subjects of the Nineties' poets, and of their successors in individualism, the Georgians, were not believed in and not familiar subjects. They were served up because they were poetical (Samarcand or linnets);[12] they had none of that aura of problem which attends so much of the best poetry. Poetry was comparatively popular because it was dope, though a mild and fairly harmless dope like cigarette-smoking. Samarcand and the linnets were consolations for the arm-chair. So, though superficially very different, were the poems of the Imagists. Imagism was a branch from the stump of Pure Form—a salutary movement in that it

[10] Siegfried L. Sassoon (1887–1967), a notable predecessor of MacNeice at Marlborough College, satirical antiwar poet widely known for public affirmation of pacifism after winning the Military Cross, the first editor of Wilfred Owen's *Poems* (1920), chiefly a memoirist after 1928.

[11] T. S. Eliot, 'The Metaphysical Poets' (1921), *Selected Essays*, 289; Eliot's italics.

[12] Examples of self-conscious, romantic diction used by Georgian poets in Sir Edward Marsh's anthologies 1911–22. Samarkand was originally evoked by Marlowe, *2 Tamburlaine* IV. i. 105 and IV. iii. 107, 130, and by Keats, 'The Eve of Saint Agnes', st. 30; later by James Elroy Flecker (1884–1915) in *The Golden Road to Samarkand* (1913) ('Santorin' in *Georgian Poetry 1913–1915* (1915), 107). Linnets had been celebrated by Richard Lovelace, 'To Althea from Prison', st. 3, and by Tennyson, *In Memoriam*, xxi, st. 6; later by Walter de la Mare (1873–1956), 'The Linnet' (1918) in *Georgian Poetry 1918–1919* (1920), 42.

insisted on clarity and precision, so reacting from the woolly or amethystine vagueness[13] of the other escapists, but itself escapist and bad in that the imagists had nothing to say. One must be clear and precise *about something*.

Then came Mr T. S. Eliot—the arch-highbrow. But the arch-highbrow writing down honestly his own view of the world is a more human, 'popular', and valuable person than the purveyor of 'poetic' subjects to a public which buys them because they are the accepted things—just as they once bought antimacassars and now buy chromium-plated clocks with unreadable dials. Mr Eliot was, indeed, extraordinarily (pathologically?) interested in literature, but he never fixed a great gulf between the street and the classics; he saw them in inter-relation.

Mr Eliot brought back into English poetry precision—the blade which the Imagists had sharpened but never used. For Mr Eliot had a subject. This is too often forgotten, and he is thought of sometimes mainly as a technical experimenter. Mr Eliot's eccentricities of form were, however, largely evoked by and appropriate to his subject. The fragmentariness of *The Waste Land* can be compared with the consonantal rhymes of Owen, the rhythms of Hopkins, the halting rhythms and bad rhymes of *Hugh Selwyn Mauberley*, and the at first sight stilted rhymes of Spender.[14] All these are appropriate to their subject.

We must not judge a writer exclusively by his form or exclusively by his subject. In considering subjects the ordinary critic is too apt to deduce that a man who writes, say, about morbid subjects is a morbid writer. The intellectual critic, on the other hand, very often denies that distinctions in subject-matter affect the values of the work (e.g. Mr Robert Graves in his early book

[13] MacNeice's coinage for 'shadowy-purple verse' by analogy with *purpureus pannus* in Horace's *Ars Poetica* (*Epistles* II. iii. 14–19)—'purple patch' of florid irrelevancy in writing.

[14] Wilfred Owen (1893–1918) constructed entire poems out of half-rhymes by varying a vowel within a box of unvarying consonants. G. M. Hopkins (1844–89) based many poems on sprung rhythm by counting stresses in a line and discounting unstressed or slack syllables. In *Hugh Selwyn Mauberley* (1920) Ezra Pound explored new possibilities against loose free verse by writing restricted strophes while introducing into the tight form syncopation and inexact rhyme on English and foreign words. Sir Stephen Spender wrote deliberate rhythmic hesitancies into his *Poems* (1933, 1934) to convey idioms of unusual candour.

of criticism, *Poetic Unreason*).[15] It is perhaps safe to say that concentration on a narrow sphere of subjects is a good, though by no means an infallible, clue to a man's spiritual outlook. Consider Mr Eliot's subject-matter. Until recently in his professedly Christian works (and possibly still in these?) his subject was the doom or (what was worse in that it lacks the grandeur of doom) the decay of our modern civilization. He treated this doom and decay frankly from the angle of the highbrow. 'Prufrock' and 'A Portrait of a Lady' were witty, shrewd, sincere, and eminently gloomy pictures of sophisticated society. Like the characters in Shakespeare whose glorifications of suicide betray their lust for life,[16] Mr Eliot's satire of his world in these earlier poems betrays that it was *his* world and the only world congenial to him. Thus Mr E. M. Forster tells of how he first read Eliot's poems during the War and found them

> innocent of public-spiritness: they sang of private disgust and diffidence, and of people who seemed genuine because they were unattractive or weak. . . . Here was a protest, and a feeble one, and the more congenial for being feeble. For what, in that world of gigantic horror, was tolerable except the slighter gestures of dissent? . . . He who could turn aside to complain of ladies and drawing-rooms preserved a tiny drop of our self-respect, he carried on the human heritage.[17]

Mr Eliot, like Henry James, saw his ladies and drawing-rooms against a dwarfing background of cultural tradition. But he also saw the modern City against the background of European history, in particular of its religious history. He passed from private disgust and diffidence to that great objectification of despair and collapse which is *The Waste Land*.

> These fragments I have shored against my ruins.[18]

[15] Graves in *Poetic Unreason and Other Studies* (1925, 1968) gave a subjective and relativistic reading of poetry from nursery rhymes and journalistic verse to Milton and Shakespeare. MacNeice criticized this 'psychological approach to poetry' taken by Graves in *Poetic Unreason* in his Oxford paper 'We are the Old' (1930): see his *Modern Poetry*, 73.

[16] Shakespeare's suicides include Romeo and Juliet, Brutus and Cassius, Othello, Antony and Cleopatra.

[17] Forster, 'T. S. Eliot' (1928), *Abinger Harvest* (1936), 87–8; the same passage quoted again in *Modern Poetry*, 12.

[18] *The Waste Land*, v. 430, taken as the key-line in *Modern Poetry*, 166.

These fragments are fragments of Christianity and Paganism, of desire for system *à la Dante*, of the lust for power which finds expression in knowledge, of the individualist Urge of Whitman, the imperial Will of Kipling, of all the ideals of sex from the Troubadours to Lawrence, of the memory of rituals and heroes, of the hope for salvation through aloofness, of the 'boredom and the glory' of London.[19]

The Waste Land is not a satirical poem, for in it Mr Eliot has passed beyond satire. The satirist is a kind of escapist. The cross-currents, switch-overs, throw-backs, and quasi-automatic tags of *The Waste Land* are a serious and horrified attempt to represent (without satire and without sentimentality) the gloomy cross-currents, the half-exposed strata, the ruins and dying roots of that civilization which the Freudians spend their time dissecting and which the Marxists hope to alter. Compare the words of Freud—

> The fateful question of the human species seems to me to be whether and to what extent the cultural process developed in it will succeed in mastering the derangements of communal life caused by the human instinct of aggression and self-destruction. In this connexion, perhaps the phase through which we are at this moment passing deserves special interest. Men have brought their powers of subduing the forces of nature to such a pitch that by using them they could now very easily exterminate one another to the last man. They know this—hence arises a great part of their current unrest, their dejection, their mood of apprehension.[20]

Such poetry as was produced by the War itself brought back with it out of chaos the values of humanity. Pity reappeared in English poetry. 'The poetry is in the pity', said Wilfred Owen, the greatest of the war poets.[21] The nineteen-thirty school of English poets, represented by Mr Auden and Mr Spender, derives largely from Owen.[22] When Auden says 'Man must unlearn

[19] Eliot, 'Matthew Arnold', *The Use of Poetry and the Use of Criticism* (1933), 106: 'But the essential advantage for a poet is not to have a beautiful world with which to deal: it is to be able to see beneath both beauty and ugliness; to see the boredom, and the horror, and the glory.'

[20] Concluding para. to Sigmund Freud's *Civilization and its Discontents*, 1st auth. English trans. Joan Rivière (1930), International Psycho-Analytical Library, no. 17, 143–4.

[21] Owen's famous Preface to his *Poems* ; cf. *Modern Poetry*, 78–9.

[22] See W. H. Auden, 'As It Seemed to Us' (1965) in *Forewords and Afterwords* (1973), 523; Stephen Spender, *World Within World* (1951, 1966), 50; Spender, *The Thirties and After* (1978), 18: 'what they wrote was anti-fascist . . . pacifist poetry', 'profoundly influenced by the diction and attitude of Wilfred Owen'.

hatred and learn love' he is affirming the values of such a poem as 'Strange Meeting';[23] compare also Spender's poem, 'The Prisoners',[24] with Owen's 'Insensibility'. These new poets have introduced an element of thrustful (sometimes even optimistic) preaching into their works, which was not possible to the war poets and which is derived partly from Communism and partly from D. H. Lawrence.[25] The pity of Owen, the Whitmanesque lust for life of Lawrence, and the dogmas of Lenin[26] are now combining to make possible the most vital poetry seen in English for a long time.

Before discussing Auden and Spender in more detail I will take the instructive example of Mr Yeats, a poet much older than Mr Eliot, who started in the nineties as an escapist like the other poets of that time and has worked his way, by devious routes of hoodoo and wilful creeds, to a poetry which is concerned with life, a limited life[27] but not so limited as Mr Eliot's and one which is of value and interest to many. Mr Yeats, like Mr Eliot, is not a poet to imitate. He has gone too roundabout to his end.[28] Auden writes in an essay on 'Psychology and Art': 'There must always be two kinds of art, escape-art, for man needs escape as he needs food and deep sleep, and parable-art, that art which shall teach man to unlearn hatred and learn love.'[29] Speaking very approximately I would say that the difference between Yeats and Eliot on the one hand and Auden and Spender on the other

[23] Conclusion to Auden, 'Psychology and Art To-day' (1935), *The English Auden*, 341–2; see below with n. 29. Owen's 'Strange Meeting' was placed first in his published *Poems* (1920).

[24] See Spender, *Collected Poems* (1955), 26–7, incl. line 'My pity moves amongst them like a breeze'.

[25] 'We are perhaps likely to forget how completely the political revolution [Communism] was, at this time, fused for most intellectuals with a vaguely Lawrentian moral revolution': Monroe K. Spears, *The Poetry of W. H. Auden*, 83–4. See Auden's 'A Communist to Others' (1932) in Michael Roberts's *New Country* (1933), not collected until *The English Auden* (1977), 120–3.

[26] The moral imperative common to Owen, Lawrence, and Lenin appears to be an admonitory comradeship, as extolled by Roberts in his anthologies *New Signatures* and *New Country* (1932; 1933).

[27] Cf. Vachel Lindsay's poem 'The Congo' (1914): 'Mumbo-Jumbo will hoo-doo you.' In his *The Poetry of W. B. Yeats* (1941, 1967), 23–5, MacNeice again attempted to show Yeats's limitations as a 'mystical' poet.

[28] Cf. the title of MacNeice's novel *Roundabout Way* (pseud. 'Louis Malone', 1932).

[29] (1935), *The English Auden*, 341–2: see n. 23 above.

is that escape-art predominates in Yeats and Eliot while parable-art predominates in Auden and Spender, though these two latter are of course themselves escaping at the same time as they are preaching parables.[30] Auden, for example, is preaching communism, psycho-analysis, and the love of one's fellows, but while and through preaching those things he will escape into a rather crude hero-worship, into schoolboy spite, nostalgia for bleak spaces, and (like Mr Eliot) into a masochistic delight in desolation and nightmare.

Mr Yeats in his earlier days not only practised but advocated escape. In an essay on 'The Symbolism of Poetry' (1900) he wrote:

> With this return to imagination, this understanding that the laws of art, which are the hidden laws of the world, can alone bind the imagination, would come a change of style, and we would cast out of serious poetry those energetic rhythms, as of a man running, which are the invention of the will with its eyes always on something to be done or undone; and we would seek out those wavering, meditative, organic rhythms, which are the embodiment of the imagination that neither desires nor hates, because it has done with time.[31]

But the later Yeats seems to have ceased to abnegate desire and hatred—'I study hatred with great diligence'[32]—and his rhythms no longer waver so much. As a metaphysician he seems to have recognized the necessity of the descent into time of 'desecration and the lover's night'.[33]

In spite of all his preoccupation with style and certain stage-room trappings which he still affects, Yeats is a salutary influence on modern poetry just because he is *not* too literary (he is less literary than Mr Eliot and far less literary than Mr Pound). He long ago saw clearly what he was trying to do, for in an early essay he tells how as a young man he made a great discovery. The discovery was this:

> We should write out our own thoughts in as nearly as possible the language we thought them in, as though in a letter to an intimate friend.

[30] MacNeice treated this topic in *Modern Poetry*, *passim*, and in *Varieties of Parable* (1965).

[31] In *Ideas of Good and Evil* (1903), collected in *Essays and Introductions*, 163; same passage quoted in *Modern Poetry*, 22, with four inserted comments.

[32] 'Ribh considers Christian Love insufficient' (1934), st. 1.

[33] See review 'The Newest Yeats' (Feb.–March 1936) above, n. 4.

We should not disguise them in any way; for our lives give them force as the lives of people in plays give force to their words. . . . 'If I can be sincere and make my language natural, and without becoming discursive, like a novelist, and so indiscreet and prosaic,' I said to myself, 'I shall, if good luck or bad luck make my life interesting, be a great poet; for it will be no longer a matter of literature at all.' Yet when I re-read those early poems which gave me so much trouble, I find little but romantic convention, unconscious drama. *It is so many years before one can believe enough in what one feels even to know what the feeling is.*[34] [Italics mine.]

It has taken Mr Yeats a long time, I think, to find out what his feelings are, and perhaps even now he sometimes deceives himself. He believes, as he says in *Dramatis Personae*, in a discipline imposed from within, not from without; a man should make his own mask.[35] But the fact that he sees the world through a series of eccentric home-made frames (theosophy, spiritualism, a mystical dialectic, aristocratic idealism)[36] does not mean that he sees it false; if there is a glass in those frames to save him from the winds which afflicted D. H. Lawrence,[37] it is only plain glass, not stained or frosted. We must not be discouraged in Yeats by his self-stylization. Dr Johnson, who had a hard head, yet suffered from neuroses, one of which (I have forgotten the exact details) was something like this:[38] whenever he had to pass through a door he would start counting in the attempt to cross the threshold on, say, the thirteenth step and on the right foot. To have such rules in life is awkward, but I can imagine that some one thus afflicted, who had a flair for elegant deportment, might conceivably turn such a tic to artistic account, always entering doors with a noticeably graceful rhythm so that people might even come to speak of him as The Man who goes through Doors So

[34] *Reveries over Childhood and Youth* (1916), xxx, in *Autobiographies* (1955), 102–3; quoted again in *Modern Poetry*, 23.

[35] *Estrangement* 1909 (1926), vii and xxii, first collected in *Dramatis Personae* (1936), later in *Autobiographies* (1955), 464, 469. Sect. xxii is quoted in *Modern Poetry*, 24; see also *Modern Poetry*, 168.

[36] Yeats's combination of esoteric pursuits is given a biographical background in *Modern Poetry*, 80–3, a fuller critical approach in *The Poetry of W. B. Yeats*.

[37] MacNeice explained the comparison between Yeats and Lawrence in *The Poetry of Yeats*, 187–8.

[38] *Boswell's Life of Johnson*, ed. G. B. Hill, rev. L. F. Powell, 6 vols. (1934–50), i. 484–5; also Miss [Frances] Reynolds, *Recollections of Dr Johnson*, in *Johnsonian Miscellanies*, ed. G. B. Hill, 2 vols. (1897), ii. 275.

Beautifully.[39] Such a one is Mr Yeats; such perhaps are all poets to some extent.

The quotation from Yeats above shows that he always recognized the importance of subject—that he falls into line, in fact, with such poets as Wordsworth and Auden.[40] It was only, however, in his later poems that Yeats began to treat the contemporary subject; witness that magnificent poem 'Easter 1916'.[41] His earlier 'Celtic' works must not, however, be regarded as purely decorative; the *Idylls of the King* and Mr Ezra Pound's *Cantos* approach much nearer to pure decoration.[42] Cathleen ni Houlihan may indeed be open to ridicule, may deserve the taunts showered upon her by James Joyce and Denis Johnston,[43] but the mere fact that she has received these taunts proves that she is a live conception which means something to many people. To an Irishman at any rate she means far more than Tennyson's Queen Guinevere could have meant to Tennyson or to any other Englishman. The Irish tend to maternal fixation, and this becomes sublimated into their peculiarly violent and sentimental patriotism. Much of Yeats's verse is a memorable expression of this emotion. Yeats is therefore nearer the ordinary man (or *some* ordinary men, i.e. Irishmen) than Eliot was until he too started writing for a group of ordinary men (i.e. English High Church Protestants).

Yeats's Celtic Twilight poems were therefore far less an escape from life than, say, the ballads of Dante Gabriel Rossetti or of William Morris,[44] poets who still influence his manner; and, strange though it may seem, all these outpourings about Cuchulain

[39] An example of MacNeice's Shavian wit concerning Yeats as an aesthetic didact like Walter Pater. Cf. Shaw's characterization of Eugene Marchbanks the poet-lover in *Candida* 1895, Act I—the second of *Plays Pleasant* (1898).

[40] Like Wordsworth and Auden Yeats wrote songs and ballads on public subjects in popular spoken idioms: e.g., the recent 'Parnell's Funeral' and 'Three Songs to the Same Tune' in *A Full Moon in March* (1935).

[41] The opening section of this poem (collected 1921) is quoted in *Modern Poetry*, 83.

[42] Medieval properties and archaic diction were used in both Tennyson's Arthurian cycle (1842–82) and Pound's *XXX Cantos* (1930, 1933).

[43] Yeats's early play *Cathleen ni Houlihan* (1902) introducing 'The Poor Old Woman' of Ireland personified was mocked in the old dairy woman of Joyce's *Ulysses* (1922), episode i, and in the old flower woman of Denis W. Johnston's *The Old Lady Says 'No'* 1929 (1932).

[44] For influence of Rossetti and Morris on Yeats's balladry, see MacNeice's article 'Poetry To-day' (Sept. 1935) above with n. 88.

and Maeve[45] were more representative of people's real feelings and outlooks than either the works of the French Symbolists or the mass of Georgian nature poetry, all those little lyrics which came forward with a simper as if to say, 'How simple we are! How true to the heart of the ordinary simple man!' Yeats and Eliot broke away from the stifling fashion of corner-poetry. 'We live in our own corner', the poets had been saying, 'apart from the rest of the room.' It did not occur to them that there is no such thing as a corner in abstraction from a room.

Still, a corner can be more or less screened off. Yeats and Eliot themselves are still fairly well screened when compared with Wilfred Owen or even with Auden and Spender. Auden and Spender have the great gift of compromise; they do not try to be purists. The purists are the curse of the arts.[46] Let us take two contemporary types as examples:

(1) *The Surrealists.* Surrealism is the belated putting into practice of the theory of poetry found in Plato's *Ion* (no wonder Plato disapproved of poetry).[47] The surrealist does not (or says he does not) control his medium in the very least. He is, in the words of one of them, just 'a modest registering machine'.[48] Now, if the surrealists could confine themselves to being modest registering machines they would presumably do useful spade-work for the psychologists. But I have not enough faith in the modesty of human beings to believe that the surrealists work within their

[45] Cuchulain, the great Irish legendary hero of Ulster, and Maeve, the great queen and his nemesis, were celebrated by Lady Augusta Gregory in two collections, *Cuchulain of Muirthemne* (1902) and *Gods and Fighting Men* (1904), both introduced by Yeats, who used the stories for his five Cuchulain plays 1904–39.

[46] Cf. Preface to *Modern Poetry*: 'This book is a plea for *impure* poetry' (MacNeice's italics), and n. 7 above.

[47] Surrealism, a 'resistance movement' in painting and poetry, begun in France 1924 by André Breton (1896–1966) and drawing upon the subconscious (incl. dreams and hallucinations) as 'the true means' of knowledge, may be related to the doctrine of inspiration found in Plato's *Ion* ; in a later dialogue, *The Republic*, Plato banned poets as liars from his ideal state. For the *Ion*, cf. *Modern Poetry*, 63.

[48] Surrealists 'have been content to be the silent receptacles of many echoes, modest *registering machines* that are not hypnotised by the pattern that they trace': André Breton, first *Manifeste du surréalisme* (1924) as quoted and translated by David Gascoyne, *A Short Survey of Surrealism* (1935), 63, and again in Breton's *What Is Surrealism?* (1934, trans. Gascoyne, 1936, 1974), 61 (Breton's italics); quoted again in *Modern Poetry*, 156. Cf. Eliot's idea of the poet as a medium or catalyst in 'Tradition and the Individual Talent' (1919), *Selected Essays*, 18: 'The mind of the poet is the shred of platinum.'

own limits. I believe that they are almost bound to exercise selection (which is on their premisses a crime), e.g. if a surrealist, while dutifully not thinking what he was writing, were modestly to register a perfectly logical syllogism, I fancy he would suppress it.

The First Manifesto of Surrealism (1924) contains this passage:

> Surrealism, as I envisage it, displays our complete *nonconformity* so clearly that there can be no question of claiming it as witness when the real world comes up for trial. On the contrary, it can but testify to the complete state of distraction to which we hope to attain here below. . . . Surrealism is the 'invisible ray' that shall enable us one day to overcome our enemies. . . . This summer the roses are blue; the wood is made of glass. The earth wrapped in its foliage makes as little effect on me as a ghost. Living and ceasing to live are imaginary solutions. Existence is elsewhere.[49]

If existence is elsewhere, art is elsewhere also. Schizophrenia and paranoia may be, as the Freudians suggest, parallel to and alternative to art;[50] they are not identical with it. The Unconscious undoubtedly has a great say in poetry, as most good critics have always recognized,[51] but, whether you call it the Unconscious or Inspiration, it is a mistake for the poet to sit back in cold blood and ask it to do all the work. It is not fair to the Unconscious. If the Unconscious is to be given a chance, the writer should be concentrating his mind on his manifest subject. It seems to me that Wordsworth's poem—

> My horse moved on; hoof after hoof
> He raised and never stopped:
> When down behind the cottage roof,
> At once, the bright moon dropped—[52]

has all the virtues (which are real virtues) of the direct statement, while at the same time it beats the surrealists off their own ground. The surface mind should deal with the surface pattern; like to

[49] Conclusion to Breton's first *Manifeste du surréalisme* as quoted in Gascoyne's *Short Survey*, 66–7, and again in Breton's *What Is Surrealism?*, trans. Gascoyne, 65 (Breton's italics); quoted again in *Modern Poetry*, loc. cit., 157.

[50] Cf. Auden, 'Psychology and Art To-day' (1935), *The English Auden*, 335. The two opening sentences of this para. are used in *Modern Poetry*, loc. cit.

[51] See MacNeice's article 'Sir Thomas Malory' (April 1936) above, n. 2.

[52] 'Strange fits of passion I have known' (1800), st. 6.

like. We must never forget that poetry is made with words, that words are primarily for communication, that verbal communication is, if you like, a surface ritual. It may not go as deep as other forms of communication, physical or religious, but, if a poet is to write at all, he should be content with the results proper to poetry. If he wants yoga he can go elsewhere.

(2) Surrealism is an extreme fashion, in its true form being psychic automatism. At the opposite pole stands the poetry of pure propaganda, such as was produced in Russia after the Revolution. The communist poet Maiakovski[53] (I quote from René Fülop-Miller's book, *The Mind and Face of Bolshevism*) 'arrived at the point of carrying on poetry purely as a trade; he proudly proclaimed that he had established a "word workshop" and was in a position to supply every revolutionary, "promptly and on easy terms", with any quantity of poetry desired'.[54] This attitude destroys the distinction between poetry and prose and undermines poetry's claim to a higher or, at least, a different kind of truth. An advertisement is not expected to be true.

Poets to-day are seduced into bill-plastering on the one hand or visceral parrot-talk on the other.[55] The Auden-Spender school of poets upholds the English tradition of freedom in that it walks a middle course. In spite of the many siren calls to perfection or profundity, these poets have the nerve to continue writing a poetry which is in three respects old-fashioned:—(1) it is in a tradition and does not attempt to be over-revolutionary in form; (2) it has a mixed content and contains nearly all the elements which other schools severally ban; (3) it is not directed solely either to entertainment or instruction, uplift or aloofness; it develops itself instinctively but with a reasonable amount of self-consciousness and self-criticism.

Mr Yeats, as shown above, proposed at one time to turn his back on desire and hatred. Mr Eliot sat back and watched other people's emotions with ennui and an ironical pity:

[53] Vladimir V. Mayakovsky (1893–1930), the individualist Russian poet and dramatist, leader of the futurist school and chief poet of the revolution, who turned to collectivist or propaganda verse.

[54] *The Mind and Face of Bolshevism*, trans. F. S. Flint and D. F. Tait (1927, 1965), 160. MacNeice has here deleted the phrase 'on receipt of a simple order' before '"promptly and on easy terms"'.

[55] See MacNeice's 'Poetry To-day' (Sept. 1935) above, n. 5.

I keep my countenance,
I remain self-possessed
Except when a street-piano, mechanical and tired
Reiterates some worn-out common song
With the smell of hyacinths across the garden
Recalling things that other people have desired.[56]
[Italics mine.]

The young poets, on the contrary, are almost blatant in their readiness to hate and love. They are essentially young poets (and young does not mean immature), whereas Yeats and Eliot even in their twenties had an autumnal quality like Tennyson and the Pre-Raphaelites. Mr Auden, the chief innovator of these poets, is a difficult type to define. He is not an emotional poet, a lyrist, in the traditional sense *à la* Sappho or Shelley, but it would be misleading to call him a poet of ideas. Though he is not a philosopher, Auden is excited not so much by individual people or things as by people or things *qua* parts of an up-to-date scheme. He sees the world in terms of psycho-analysis or in terms of a Marxian doctrine of progress; these overlap and to some extent co-operate. The great advantage for Auden in believing in these two doctrines severally or in his private blend of both is that thereby nearly all the detail in the world becomes significant. For Auden *qua* psychologist nearly anything henceforward will be either (*a*) an example or symbol of a neurosis which needs curing *à la* Freud, or (*b*) an example or symbol of how a neurosis produces good (for Auden believes that all progress is due to neurosis). For Auden *qua* Marxist, on the other hand, nearly anything will be either (*a*) a product of the Enemy,[57] the reactionary, bad and therefore to be fought against, or (*b*) a relic of the past, once perhaps good but now bad and to be deplored, though often with reverence and affection, or (*c*) an earnest of better things, a pioneer of the future—or else a symbol of one of these three types. It will be seen that the neurotic who needs curing will often be identifiable with the political enemy, while the productive neurotics will include the political reformers.

[56] 'Portrait of a Lady', ii. 37–42.

[57] For the Enemy in Auden's early work, see John Fuller, *A Reader's Guide to W. H. Auden* (1970), 54, 64, 68–9.

Mr Spender, on the other hand, is primarily a *feeling* poet. He is not a born missionary like Auden. Auden sees a stranger and is excited because the stranger is either a co-missioner or an enemy or (in most cases) a possible convert. His reaction, therefore, is not logically immediate because it always contains an ideological factor. Spender, on the other hand, reacts immediately, instinctively, and tends to tack his ideology on afterwards.

It is sometimes objected against these younger poets that their 'modern' stage-properties are a little obvious; that they introduce pylons and gasometers as automatically as older poets introduced roses and nightingales. This is often true, but it should be remembered that pylons and gasometers are not merely décor. The modern poet is very conscious that he is writing in and of an industrial epoch and that what expresses itself visibly in pylons and gasometers is the same force that causes the discontent and discomfort of the modern individual, the class-warfare of modern society, and wars between nations in the modern world.

Mr Yeats in his excellent 'Dialogue of Self and Soul' seemed on the whole to be gambling on the Self.[58] It is the Self which plunges in and swims the world of appearance; the Soul sits apart on an ice-floe like a yogi. Mr Eliot puts all his hopes on the Soul. He would agree with Yeats that

> Love has pitched his mansion in
> The place of excrement,[59]

but whereas Yeats in his later poems declaims this defiantly and exultingly, Eliot's cloacal lovers are not allowed to find a Salvation *here and now* through any dialectic of ethical or æsthetic opposites; their only hope is to turn their back on this world, a world which, for Eliot, is an 'eructation of unhealthy souls'.[60] Yeats would imply that, if the world is a belch, it is the valiant belch of a Gargantua—

> and all these I set
> For emblems of the day against the tower

[58] This poem (1933) opens with a balance of forces, self and soul, but closes with an assertion of self alone.

[59] 'Crazy Jane Talks with the Bishop' (1932–3), st. 3.

[60] *Burnt Norton* (1935), iii. 19.

> Emblematical of the night,
> And claim as by a soldier's right
> A charter to commit the crime once more.[61]

Auden and Spender follow Yeats rather than Eliot in being more concerned with the Self than the Soul. They recognize the existence of the Waste Land but believe that its fertility will be restored. This is especially obvious in Mr Day-Lewis, whose hopes of a new world are perhaps a little too orthodox for a proper fusion in poetry.[62] I find the note of hope, or at least of the will to live, in even such a gloomy poem as the last chorus of Auden's 'Paid on Both Sides'—

> Though he believe it, no man is strong.
> He thinks to be called the fortunate,
> To bring home a wife, to live long.
>
> But he is defeated; let the son
> Sell the farm lest the mountain fall;
> His mother and her mother won.
>
> His fields are used up where the moles visit,
> The contours worn flat; if there show
> Passage for water he will miss it;
>
> Give up his breath, his woman, his team;
> No life to touch, though later there be
> Big fruit, eagles above the stream.[63]

This poem is far more 'Greek' than anything in A. E. Housman. By 'Greek' I mean that it is the economical expression of an emotion which is not egocentric. Epicureanism is egocentric and therefore lacks tension.[64] Auden's admiration for the objective world is founded in that cosmic pride which is distinct from personal pride and which is at the base of Christianity. This explains his belief that 'Pelmanism'[65] is an important factor both in art and in the good life. The Epicurean, like the Artist for Art's Sake, will have no use for pelmanism. Why burden his mind with

[61] 'Dialogue of Self and Soul', i, st. 4.
[62] Day-Lewis' poetry of the 1930s had become overt communist propaganda.
[63] 1928 (1930), conclusion: *The English Auden*, 17.
[64] Statement repeated in *Modern Poetry*, 87.
[65] Memory training by methods of the Pelman Institute; see *Modern Poetry*, 87–8.

facts *which cannot affect his own life*? The Epicurean does not appreciate Otherness as such.[66]

Spender, too, has remarked in conversation that he believes in 'touchability'. Touchability is, on the emotional or physical plane, the counterpart of Auden's pelmanism on the intellectual. It means the renunciation of the utilitarian Epicurean self and the belief that people in themselves are worth knowing and touching, just as for Auden facts are worth remembering.[67] The Epicurean lover, strictly speaking, is only interested in what he can get out of love; his love is a private luxury, his beloved a temporary piece of furniture. Many love-poets have taken this attitude. Spender is a love-poet on a higher plane; like a true Christian he recognizes that human beings are ends in themselves and must not be degraded into tapestried figures, into hearth-rug pets, harem favourites, or valets.

Auden and Spender have reasserted the truth which was expressed, though badly, in Wordsworth's introduction to the *Lyrical Ballads*.[68] Every man lives in a contemporary context which is of value and interest. That is the life which, directly or indirectly, he should write about. He must be a craftsman to be able to 'put it over', and this he may have to do under various disguises. Auden and Spender are good craftsmen, as are Yeats and Eliot. This essay has been concerned with subject, but I would add, though it is a truism, that a verbal sense is as necessary for poetry as a good eye is for ball games. When a poet has a verbal sense but little to say, he wastes himself in the sands like Mr Pound.[69] It is a good thing that there are poets now writing who have as much to say and can say it as well as those whom I have been discussing.

[66] Awareness of Otherness is essential to interpersonal life: *Modern Poetry*, 88.

[67] The three statements opening this para. were repeated in *Modern Poetry*, loc. cit.

[68] Wordsworth's Preface (1800, as rev. 1802) to *Lyrical Ballads* incorporates a passage on the question 'What is a Poet?' incl. the famous answer, 'He is a man speaking to men.' The passage continues in self-congratulatory terms. Cf. MacNeice's preference for 'poets whose worlds are not too esoteric' in *Modern Poetry*, 198.

[69] MacNeice considered Pound a dilettantish poet compared to Eliot: *Modern Poetry*, 85.

Look, Stranger! Poems. By W. H. Auden[1]

Look, Stranger! Poems by W. H. Auden (Faber. 5*s*.)

Mr Auden is a poet who keeps developing. Development usually means a certain loss and these poems do not show the fascinating accumulation of tense, if difficult, phrases, the beautiful telegraphese which marked his first collection. But the gain is far greater than the loss (contrast Mr Eliot, who also has been moving towards the direct statement). Mr Auden is a missionary but, unlike many missionaries, he has an eye. It is the eye which keeps the balance between emotion and intelligence, between Shelley and Eliot. Escapism is extremism. That criticism of life which is the function of major poetry (of what Mr Auden himself calls parable-art)[2] is the product of writers who, however much they may take sides, or however much they may rationalize, yet manage to remain in contact with what, on the analogy of the concrete universal,[3] we may call the incarnate problem—i.e., human nature. For it is Mr Auden who has brought back humanity into English poetry.

But Mr Auden is not only serious, he is entertaining (which only reminds us how boring is pure frivolity). He gives us lavishly the pleasures which are proper to poetry. This book, though homogeneous, is admirably varied. There are, of course, many sermons launched over the English landscape (compare the declamatory choruses of *The Dog Beneath the Skin*). But there are also what people would call 'pure lyrics' and sonnets reflective or passionate. Then there is the delightful decorative exercise *à la* Sir Philip Sidney beginning 'Hearing of harvests rotting in the

[1] Unsigned review in the *Listener* 16: 416 (30 Dec. 1936), 1257. Auden was awarded the King's Gold Medal for poetry for this book which, in the American edition, was entitled *On This Island* (1937).

[2] For Arnold's 'criticism of life' see 'Poetry To-day' (Sept. 1935) above with n. 8; for escape-art and parable-art see Auden, conclusion to 'Psychology and Art To-day' (1935), *The English Auden*, 341–2.

[3] 'Concrete universal': term in English Hegelian philosophy from J. H. Stirling (1820–1909) to F. H. Bradley (1842–1924) and Bernard Bosanquet (1848–1923), or roughly 1865–1912, explained as 'the individual, when regarded as something maintaining its identity through qualitative change or diversity': *OED*, Suppl. i (1972), 605.

valleys',[4] which is a proof that decoration and 'a message' are not, as some think, incompatibles. There is the excellent comic broadsheet beginning 'O for doors to be opened and an invite with gilded edges'.[5] There is a vigorous commination service directed at 'the great malignant Cambridge ulcer'.[6] There is a letter containing passages of unabashed allegory, and there is a puzzle poem (Number III) which shows that the old Auden is not dead yet (needless to say, the puzzle works out).[7]

It is to be noticed that Mr Auden is writing much more in stanzas and with rhyme. He still perhaps makes excessive use of false rhyme which is like those sinister colours that should not be laid on everywhere. Possibly, too, he is a little slapdash. But, being a serious artist, he is interested in his craft as a craft. It is perhaps superfluous at this stage to point out that he has a great knack with words. Or that he has more vitality than all his fellow-poets put together. What cannot be repeated too often is that he is a poet who has something to write about. This something is a mix-up of politics and psychology,[8] repellent no doubt to purists but sympathetic to that large minority who, too busy living to be thinkers, yet temper their lives with ideas. Compare the objectivity of Horace and contrast Prufrock and Mauberley with their blend of self-pity and satire.

In having something to say, Mr Auden is in the tradition of Dryden, Byron and Wordsworth rather than of Keats, Tennyson, the 'nineties poets and the Georgians. Further, he is a religious poet and it is pleasant to notice in these later poems that of the two contradictory convictions which cohabit in every religion—(1) that

[4] Sestina in *The English Auden*, 135–6—a troubadour verse form of six six-line stanzas using an interlocking rhyme-scheme of six end-words. Examples are to be found in Sidney's *Old Arcadia* (1590), iv, poems 70, 71, 76: *The Poems*, ed. William A. Ringler, Jr. (1962), 108–13, 129–30.

[5] *The English Auden*, 154–5.

[6] First published as 'A Communist to Others' in *New Signatures* (1932) with six additional stanzas; then, revised, as 'Brothers, who when the sirens roar' (1936): *The English Auden*, 120–3. 'Commination service' is an Anglican liturgical office of denunciation of sin, appointed in *The Book of Common Prayer* to be read after the Litany on Ash Wednesday and at other times.

[7] 'August for the people and their favourite islands', a birthday poem for Christopher Isherwood (Aug. 1935), *The English Auden*, 155–7, which uses allegorical personifications in st. 9–10; 'Our hunting fathers told the story', op. cit., 151.

[8] MacNeice writes more fully on Auden's political and psychological ideas in his 'Letter to W. H. Auden', Oct. 1937 (Nov. 1937) below.

the world is rotten and (2) that it is good to be alive—Mr Auden no longer puts the emphasis chiefly on the former. And if anyone still believes that poets legislate here is a poet who may suit him.

The Bradfield Greek Play: *Oedipus Tyrannus*.[1]

This is the first time that *Oedipus Tyrannus* has been acted by the boys of Bradfield.[2] It is not an easy play for boys to act, being a play of geometrical rather than realistic emotions.[3] Rhythm rather than 'acting' is what is called for. I found the necessary rhythm lacking, perhaps because the production aimed at archæological correctness rather than truth to the spirit. Truth to the spirit is a vague thing to aim at, and may be missed completely. The archæologist cannot miss completely, but are his hits worth making? I could myself do without the 'six lyres and four *auloi*, models of authentic Greek instruments, constructed in 1900, under the direction of Mr Abdy Williams.'[4] They were not good to look at, and they produced the most dismally churchy kind of noise. It must be admitted, however, that the Bradfield production has to be archæological rather than eclectic. A producer who has a miniature Greek theatre ready to hand would be unnecessarily perverse if he did not aim at a whole in keeping with it. It was a gallant experiment of Bradfield College to construct this Greek theatre[5] and whenever a play is acted in it the gallant experiment is repeated.

The grand thing about the Bradfield theatre is that the chorus can be put where they belong—in the *orchestra*.[6] The main actors are then raised decently above them, entering from the back of the stage from an alien world, while subsidiary characters

[1] The *Spectator* 158: 5687 (25 June 1937), 1187. The performance is cited in MacNeice's *I Crossed the Minch* (April 1938), 100–1. The open-air Greek Theatre at Bradfield College, Berkshire (founded 1850), was constructed in 1890: see *A History of Bradfield College*, ed. Arthur F. Leach (1900), 195.

[2] Afternoon performances on 19, 22, 24, and 26 June 1937: notice in the *Times*, 13 March 1937, 15c. This was a traditional production in the original Greek.

[3] Geometry of a complex plot dominates the play.

[4] MacNeice quotes the programme notes. *Auloi* are pipes or flutes. Charles F. Abdy Williams (1855–1923), an English composer devoted to ancient Greek music and plainsong, was director of the Bradfield Greek Theatre 1895–1901.

[5] Constructed 1890: see n. 1 above.

[6] 'Dancing space' for the chorus.

who are half-way between the heroic world of action and the static observation of the chorus enter from the *parodoi*[7] into the *orchestra* and then mount the steps to the stage from the front. The fifteen members of the chorus hold their position *in the middle of the audience* and so act as intermediaries between the audience and the drama itself. That at least is what they should do, but I was not sure that in this case they succeeded. It is difficult to identify oneself with fifteen boys wearing improbable woolly beards and awkward robes in hideous terra-cottas and purples. While realistically unsympathetic the Bradfield chorus failed also to support the play, as they could have done, by sheer rhythm. They were grouped, but not rigorously, in set patterns and occasionally moved their hands in gestures which were an unfortunate compromise between formalism and realism. The music was also a compromise, as admitted in the programme—'This setting is neither written in the musical language of to-day, nor can it claim any association with ancient Greek tonality. The musical policy has been to imbue the material with an air of antiquity, care having been taken to avoid establishing the style of any definite period.'

It is no use pretending that it is an easy thing to produce, to act, or even to speak a Greek tragedy. The speaking of this play was on the whole admirable, though some of the actors sometimes overdid their realistic intonations. It is no mean feat to be able to speak great chunks of a dead language or passages of heart-breaking stichomythia with clarity, dignity and credibility. I must, however, criticize the acting. It ought to have been much sharper and harder. The whole play has a cold but intense austerity which at all costs must not be fluffed or muffled. Oedipus ought never to have been allowed to wring his hands—a trivial and unconvincing gesture—before going out to blind himself. Jocasta was too soft. The first messenger was ill-advised to adopt a rustic drawl, while the second messenger was too juvenile in appearance though he delivered his lines with beauty and uncommon spirit. The dramatic pauses in the middle of speeches were too frequent and not very dramatic; there was a bad one in the quarrel between Oedipus and Tiresias, Oedipus self-consciously turning his back on Tiresias, while Tiresias labours to bring his next words to birth. Oedipus himself should have been better made up and should have been played

[7] Side wings leading to the orchestra.

more formally. Playing Oedipus, a boy is less handicapped by immaturity than when playing Othello or Lear. The mere pattern of the play should carry over the character.

I enjoyed my afternoon at Bradfield, but the play itself failed to take hold of me, in spite of the solid screen of trees separating us from the world and in spite of the real fire on the altar in the *orchestra*.[8] I noticed a number of clergymen in the audience, and was sadly reminded that a Greek play, once the marriage of religion and entertainment, is now the marriage of entertainment and education.[9] This is the real reason why the *Oedipus* at Bradfield is only a gallant failure.

The Note-Books and Papers of Gerard Manley Hopkins[1]

The Note-Books and Papers of Gerard Manley Hopkins. Edited with Notes and a Preface by Humphry House. (Oxford University Press. 25*s*.)

There are two extreme views about Hopkins—(1) that his religion hampered and destroyed his poetry, (2) that his poetry was derived from his religion and would not have existed without it. (I have read an article arguing that his poetry was positively influenced by the Spiritual Exercises of St Ignatius.)[2] I prefer to think that this latter view is nearer the truth, though it could not be maintained that in Hopkins dogma and the æsthetic instinct are as happily wedded as they are in Dante; but Dante after all has few parallels. Hopkins' peculiarities of form—'sprung rhythm'[3]

[8] A sacrificial fire, reminder of a defunct ritual action.

[9] Athenian drama originated in the worship of Dionysus, the plays being performed at his festivals.

[1] *Criterion* 16: 65 (July 1937), 698–700. Hopkins's *Note-Books and Papers* (1937)—read by MacNeice after his April 1937 trip to the Hebrides (*I Crossed the Minch*, 21)—then only partially published, were re-edited complete in two volumes: *The Journals and Papers*, ed. Humphry House and Graham Storey, and *The Sermons and Devotional Writings*, ed. Christopher Devlin, SJ (both 1959). Hopkins's *Letters* appeared in three volumes (1935 and 1938; reissue 1955, 2nd edn 1956); *The Poems* in four editions, eds Robert Bridges (1918), Charles Williams (1930), W. H. Gardner (1948), Gardner and Norman H. MacKenzie (1967).

[2] Humphry House, 'A Note on Hopkins's Religious Life', *New Verse*, no. 14 (April 1935), 3–5, in the Hopkins Issue to which MacNeice contributed 'A Comment'.

[3] See Hopkins, 'Author's Preface' to *The Poems*, 4th edn corr. (1970), 47–8.

and the rest—seem to me to have been due to two causes—(1) to pure technical curiosity, (2) to psychological peculiarities which could not have found adequate expression in the more normal forms of English verse. It would need a very subtle critic to disentangle these two causes. Both need stressing—the second because some poets tend to imitate the Hopkins letter while being miles away from the Hopkins spirit,[4] the first because high-flown criticism so often ignores the fact that writers, however high-minded or other-minded, do write primarily because they like writing.[5] Beliefs and enthusiasms give a writer a heart, a mainspring. He also needs hands and an eye. Hopkins lacked a sense of humour but he had that other thing which makes life livable and writings readable—an inquisitive interest in the objective world, a passion to record it precisely. The journal which occupies a hundred odd pages of this new book edited by Mr House shows abundantly Hopkins's voracity for objects—the shapes of water, of clouds, of plants, of Gothic architecture (Hopkins admired Butterfield but with discrimination).[6] The book contains fourteen drawings which are similarly not self-expression but studies of objective intracies. Hopkins is always looking for what he calls 'inscape',[7] the factor which gives a living unity to patterns—e.g. 'caught that inscape in the horse that you see in the pediment especially and other bas-reliefs of the Parthenon and even which Sophocles had felt and expresses in two choruses of the *Oedipus Coloneus*, running on the likeness of a horse to a breaker, a wave of the sea curling over.[8] I looked at

[4] Probable allusion to the didactic Marxist verse of C. Day-Lewis, who imitated Hopkins's stylistic devices without regard for their psychic or spiritual sources; Auden, Spender, and MacNeice were also influenced by Hopkins's style.

[5] MacNeice clarified what makes writers write in *Modern Poetry*, 31–5, 197–201.

[6] For Hopkins on Butterfield see his Journal, 12 June and 18 August 1874: *Note-Books*, 197 and 206; *Journals*, 248 and 254–5. William Butterfield (1814–1900), leading architect of the Gothic Revival, had carried out a major restoration of Merton College Chapel 1849–50 (Merton was the college of George Saintsbury, H. W. Garrod, T. S. Eliot, and MacNeice himself): see *Journals*, 330 (n. 4). At Oxford MacNeice had 'deplored the absurd buildings by Butterfield in Christ Church [Auden's college] and Merton': *The Strings are False*, 108.

[7] 'Inscape'—a term first used by Hopkins in his Oxford notes 'Parmenides' (1868), *Note-Books*, 98, and *Journals*, 127, and often referred to by him—is very adequately expounded here by MacNeice; see subject-index to *Note-Books* or *Journals* under 'inscape'. The term has continued to haunt criticism of Hopkins as a key to understanding his art and thought.

[8] *Oedipus Coloneus* 707–19, as noted in *Note-Books*, 391; *Journals*, 429.

the groin or the flank and saw how the set of the hair symmetrically flowed outwards from it to all parts of the body, so that, following that one may inscape the whole beast very simply.'[9] Hopkins' descriptions of clouds and water I often find very difficult to translate into pictures but his zeal in recording the visions of his bodily eye I find extremely refreshing and salutary—'Standing on the glacier saw the prismatic colours in the clouds, and worth saying what sort of clouds: it was fine shapeless skins of fretted make, fall of eyebrows or like linings of curled leaves which one finds in shelved corners of a wood.'[10] Hopkins was a believer-poet in a period of dilettantes and negativists. Since his time we have had our fill of the latter but now in the nineteen-thirties beliefs have come back into poetry with Mr Eliot and Mr Auden.[11] New brooms, however, tend to sweep out the good with the bad. Mr Eliot has thrown a spiteful word at nature-poetry[12] while Mr Auden, speaking of Wordsworth, has declared that any poet is a fool who writes about flowers.[13] While some of the art-critics are beginning to prescribe a diet strictly of sociology and to ban still lives and landscapes.[14] Perhaps this ruthless puritanism is necessary at the moment. I whole-heartedly admit that the human subject is of supreme importance; if, therefore, the painting of still lives injures the human subject, let still lives go to the scrap-heap. We might remember, however, that man is a ζῷον as well as πολιτικὸν[15] and that quite a number of people have an organic

[9] Journal, 6 April 1874: *Note-Books*, 189; *Journals*, 241–2.

[10] On the Rhône glacier: Journal, 20 July 1868: *Note-Books*, 107; *Journals*, 179.

[11] T. S. Eliot had introduced religious belief and W. H. Auden political commitment.

[12] 'Shelley was the first . . . of Nature's M.P.'s': Eliot, *The Use of Poetry and the Use of Criticism* (1933), 25–6.

[13] Auden, 'Letter to Lord Byron', iii (Aug. 1936), st. 10 and st. 12:

> I'm also glad to find I've your authority
> For finding Wordsworth a most bleak old bore. . . .
> And new plants flower from that old potato. [i.e. from Wordsworth]

—*Letters from Iceland*, 101; *The English Auden*, 183. See MacNeice's 'Letter to W. H. Auden' (Nov. 1937) below, n. 7.

[14] In 1934 'Socialist Realism' had been pronounced dogma for all Soviet artists to follow; this aimed to produce inspiring work focused on the dignity of the working man in building Communism. In 1932 the Union of Soviet Writers had proclaimed socialist realism as compulsory literary practice. Many leftist artists and writers in England were influenced in the 1930s.

[15] A 'living being/animal' as well as a 'political being/citizen': from Aristotle, *Politics* I. 2, 1253^{a} 3, 'Man is by nature a political animal'.

sympathy with trees, mountains, flowers, or with a painting by Chardin.[16] Does this sympathy lessen our sympathy with our human fellows? If so, it is a pity. It does not seem to have thus affected Hopkins.

I find the journal the most interesting part of this collection. Mr House has also put in some early poems and extracts from early diaries. I notice here a philological interest in words, more visual notes—e.g. 'Drops of rain hanging on rails etc. seen with only the lower rim lighted like nails (of fingers)'[17]—and the sympathetic remark that the fortunate losses of literature include the lost books of *The Faerie Queene*.[18] There is also an early Platonic dialogue on the Origin of Beauty[19]—rather stilted and humourless as a dialogue and disappointing as æsthetic. Hopkins typically makes the discussion hinge on the seven-leaved fan of a chestnut. Then there are notes on Parmenides and on metre and rhythm.[20] These latter do not throw as much light on Hopkins's own practice as some of the published letters to Bridges and Dixon but contain some pregnant observations—e.g. 'in everything the more remote the ratio of the parts to one another or the whole the greater the unity *if felt at all*' (italics mine).[21] The six sermons chosen by Mr House are interesting but I find them disappointing and distressing, no doubt because I am not a Catholic. Lastly there are forty pages of Comments on the Spiritual Exercises of St Ignatius Loyola which I find very difficult, no doubt for the same reason.[22] It is interesting to see that Hopkins repeats Aristotle and is inclined to believe that among animals 'the specific form, the form of the whole species, is nearer

[16] Jean-Baptiste-Siméon Chardin (1699–1779), the greatest French painter of still life until Paul Cézanne (1839–1906). Cf. MacNeice's poem 'Nature Morte' 1933, ll. 11–12: 'And in your Chardin the appalling unrest of the soul / Exudes from the dried fish and the brown jug and the bowl'—*The Collected Poems*, 21.

[17] Note, early 1866: *Note-Books*, 53; *Journals*, 72.

[18] Note, autumn 1864: *Note-Books*, 33; *Journals*, 49.

[19] In note-book dated 12 May 1865: *Note-Books*, 54–91; *Journals*, 86–114.

[20] Written 1868 and 1873, respectively: *Note-Books*, 98–102, 221–48; *Journals*, 127–30, 267–88.

[21] Lecture on rhythm: *Note-Books*, 241; *Journals*, 283. Quoted again in *Modern Poetry*, 115.

[22] MacNeice's difficulties with the Comments were probably due to the influence on Hopkins of the complex scholastic philosophy of the medieval Franciscan Duns Scotus (*c.*1266–1308), the Subtle Doctor.

being a true Self than the individual'.[23] The whole book is conscientiously annotated, and Mr House as a most diligent and sympathetic editor deserves our thanks for throwing further light on a poet unique in our literature.[24]

Letter to W. H. Auden[1]

October 21st, 1937

Dear Wystan,

I have to write you a letter in a great hurry and so it would be out of the question to try to assess your importance. I take it that you are important and, before that, that poetry itself is important. Poets are not legislators (what is an 'unacknowledged legislator' anyway?),[2] but they put facts and feelings in italics, which makes people think about them and such thinking may in the end have an outcome in action.

Poets have different methods of italicization. What are yours? What is it in your poetry which shakes people up?

It is, I take it, a freshness—sometimes of form, sometimes of content, usually of both. You are very fertile in pregnant and unusual phrases and have an aptitude for stark and compelling texture. With regard to content, the subject-matter of your poems is always interesting and it is a blessing to our generation, though one in the eye for Bloomsbury,[3] that you discharged into poetry the subject-matters of psycho-analysis, politics and economics.

[23] 'On the Principle [and] Foundation': *Note-Books*, 316; *Sermons*, 128. Cf. Aristotle, *Metaphysics* VII. 12–15, 1037^b 8–1040^b 4.

[24] Arthur Humphry House (1908–55), the distinguished editor of Hopkins, was also the author of *The Dickens World* (1941; 2nd edn 1942), *Coleridge* (1953), essays and broadcast talks *All in Due Time* (1955), lectures on *Aristotle's Poetics* (1956).

[1] *New Verse*, nos. 26–7 (Auden Double Number, Nov. 1937), 11–13. The number opens with Auden's poem 'Dover', features articles by Isherwood, Spender, MacNeice, Grigson, Edgell Rickword (1898–1982), continues with comments by sixteen contemporary writers, and closes with a bibliography of Auden's writings.

[2] MacNeice was correcting his own Shelleyan allusion to Auden as poet-legislator at the end of his review of *Look, Stranger!* (Dec. 1936) above.

[3] The Bloomsbury Group of artists and writers 1904–39 (Virginia and Leonard Woolf, Vanessa and Clive Bell, E. M. Forster, J. M. Keynes, Roger Fry, Lytton Strachey) stood for 'significant form' (Clive Bell's æsthetic term) against using 'important' ideas in art and writing as promoted by Auden.

Mr Eliot brought back ideas into poetry but he uses the ideas, say, of anthropology more academically and less humanly than you use Marx or Groddeck.[4] This is because you are always taking sides.

It may be bad taste to take sides but it is a more vital habit than the detachment of the pure æsthete. The taunt of being a schoolboy[5] (which, when in the mood, I should certainly apply myself) is itself a compliment because it implies that you expect the world and yourself to develop. This expectation inevitably seems vulgar to that bevy of second-rate sensitive minds who write in our cultured weeklies.

'Other philosopies have described the world; our business is to change it.'[6] Add that if we are not interested in changing it, there is really very little to describe. There is just an assortment of heterogeneous objects to make Pure Form out of.

You go to extremes, of course, but that is all to the good. There is still a place in the sun for the novels of Virginia Woolf, for still-life painting and for the nature-lover. But these would probably not survive if you and your like, who have no use for them, did not plump entirely for something different.

Like most poets you are limited. Your poems are strongly physical but not fastidiously physical. This is what I should expect from someone who does not like flowers in his room.[7]

Your return to a versification in more regular stanzas and rhymes is, I think, a very good thing.[8] The simple poem, however, does not always wear too well. At first sight we are very pleased to get the swing of it so easily and understand it so quickly, but after first acquaintance it sometimes grows stale. A. E. Housman,

[4] T. S. Eliot deliberately noted his use of anthropological ideas in end-notes to *The Waste Land* (1922); Auden spontaneously tossed political and psychological ideas—from Karl Marx and George W. Groddeck (1866–1934)—into his early verse.

[5] e.g., in this Auden Number of *New Verse*, Herbert Read notes in Auden 'Schoolboy jokes and undergraduate humour'.

[6] Karl Marx, *Theses on Feuerbach* (1888): 'Die Philosophen haben die Welt nur verschieden *interpretiert*; es kommt aber darauf an, sie zu *verändern*' ('Philosophers have only *interpreted* the world in different ways; what matters is *to change* it')—Marx and Friedrich Engels, *Werke*, 39 vols. (1959–68), iii. 558 (Marx's italics).

[7] A personal dislike of Auden's (*Modern Poetry*, 89; *The Strings are False*, 114), whereas MacNeice liked flowers.

[8] Auden's 'return' to regular verse forms, after Pound and Eliot's modernist free verse, was due to Hardy's example: Auden, 'A Literary Transference', *Southern Review*, 6: 1 (Hardy Centennial Issue, Summer 1940), 78–86.

whom I join you in admiring, was a virtuoso who could get away with cliché images and hymn-tune metres, but, as you would, I think, admit, his methods are not suitable to anyone who has a creed which is either profound or elaborate.

I am therefore a little doubtful about your present use of the ballad form.[9] It is very good fun but it does not seem to me to be your *natural* form as I doubt if you can put over what you want to say in it. Of course if you can put over half of what you want to say to a thousand people, that may well be better than putting over two thirds of it to a hundred people. But I hope that you will not start writing down to the crowd for, if you write down far enough, you will have to be careful to give them nothing that they don't know already and then your own end will be defeated. Compromise is necessary here, as always, in poetry.

I think you have shown great sense in not writing 'proletarian' stuff (though some reviewers, who presumably did not read your poems, have accused you of it).[10] You realize that one must write about what one knows. One may not hold the bourgeois creed, but if one knows only bourgeois one must write about them. They all after all contain the germ of their opposite. It is an excellent thing (lie quiet Ezra, Cambridge, Gordon Square,[11] with your pure images, pure cerebration, pure pattern, your scrap-albums of ornament torn eclectically from history) that you should have written poems about preparatory schools.[12] Some of the Pure Poets maintained that one could make poems out of anything, but on the ground, not that subject was important, but that it

[9] Earlier in 1937 (Jan.–Aug.) Auden wrote cabaret songs and light satiric ballads such as 'Blues', 'Johnny', 'Miss Gee', although he was to write a serious, metaphysical ballad 'As I Walked Out One Evening' (Nov. 1937): *The English Auden*, 209–10, 213–16, 218–28.

[10] i.e. in the earlier 1930s; reviewers of Auden's recent *Look, Stranger!* (1936) dealt with Auden's talent in poetic rather than political terms: G. W. Stonier in the *New Statesman and Nation*, 12: 299 (14 Nov. 1936), 776, 778; D. Powell in the *London Mercury*, 35: 205 (Nov. 1936), 76–7.

[11] Respective allusions to Ezra Pound (Canto i. 68: 'Lie quiet Dives'), F. R. Leavis's Cambridge quarterly *Scrutiny*, London centre of the Bloomsbury Group (see n. 3 above).

[12] See Auden's 'Six Odes' in *The Orators* (1932, 1934), 'I have a handsome profile' (1932), 'Out on the lawn I lie in bed' (1933), 'School children' (1937): *The English Auden*, 94–100, 123–5, 136–8, 216–17.

didn't matter.[13] You also would admit that anything can go into poetry, but the poet must first be interested in the thing in itself.

As for poetic drama, you are now swinging away from the Queer Play.[14] This, like the formal change in your lyrics, is also a good thing and also has its danger. But the danger is not so great for you as it would be for some. Whatever the shape of your work, it will always have ideas in it. Still, when authors like Denis Johnston, who can write excellent straight plays, feel impelled to go over to crooked plays and 'poetic writing',[15] there must be some good reason for it and it may appear perverse in you to forget your birthright[16] and pass them in the opposite direction to a realism which may not be much more natural to you than poetry is to them.

These are the criticisms which occur to me at the moment. I have no time to expand on your virtues, but I must say that what I especially admire in you is your unflagging curiosity about people and events. Poetry is related to the sermon and you have your penchant for preaching,[17] but it is more closely related to conversation and you, my dear, if any, are a born gossip.[18]

Yours ever,

Louis MacNeice

[13] Cf. MacNeice, Preface to *Modern Poetry*, and article 'In Defence of Vulgarity' (Dec. 1937) to be published in the second volume of selected prose.

[14] In drama, Auden's strangely inverted *Paid on Both Sides* (1930) and *The Dance of Death* (1933) were followed by popular collaborations with Christopher Isherwood: *The Dog beneath the Skin*, *The Ascent of F6*, *On the Frontier* (1935, 1936, 1938).

[15] The Irish dramatist Denis W. Johnston (1901–84) wrote both naturalistic ('straight') and expressionist ('poetic') plays, *Storm Song* and *A Bride for the Unicorn* 1933–4 (1935). See 'The Play and the Audience' (1938), below, n. 12.

[16] Auden's birthright was modernism; therefore, his early verse had been obscure and difficult. See Edward Mendelson, Preface to Auden's *Selected Poems: New Edition* (1979), xi.

[17] See the Vicar's lunatic sermon ('Depravity' (1933)), in last scene of *The Dog* (1935), first published as 'Sermon by an Armament Manufacturer' (1934): *The English Auden*, 138–41.

[18] Cf. Auden, 'In Defence of Gossip', *Listener* 18: 467 (22 Dec. 1937), 1371–2: 'Gossip is creative. All art is based on gossip—that is to say, on observing and telling.' Gossip was part of MacNeice's contribution to *Letters from Iceland* (1937), chap. 12 ('Hetty to Nancy'), 156–99.

The Play and the Audience[1]

The present-day London theatre is, on the whole, a snob institution. Country cousins inquire what shows are on in town and boys at school are told by their masters that to go to the theatre is part of one's education. But while I have met a few people who do go to the theatre in the hope of 'learning something', of getting uplift, of developing themselves, the great majority go because it is the thing to do; to go to a West End play is breaking bread with Society.

The audience with which I am best acquainted is myself. I find that I also go to the theatre for snob reasons. If I want entertainment I go to the cinema, if I want excitement I go to a rugger match, if I want wit I listen to one or two of my friends who can provide it, if I want intellectual communion I read a book. The theatre in these respects is for me an 'also ran'. I go to the theatre, on the contrary, if I want to cut a dash—to impress myself or my companions. And so I feel that half the point of the theatre is lost if I do not sit in the stalls. As with a very grand restaurant, once one gets there one may as well do the job thoroughly.

A great many people go to a play, as they might go up Snowdon or visit the Giant's Causeway, in order to say they have been there. I have studied my neighbours at West End plays in all parts of the house and, as a rule, very few of them appear to be enjoying the play itself or to be exercising their critical faculty. Many of them, however, are pretending to be critical because it will be a good mark for them later if they can say to their friends, 'Peggy Ashcroft[2] was not so good in $X^2 - Y^2$ as she was in $Y^2 - X^2$.'

[1] *Footnotes to the Theatre*, ed. R[ichard] D[enis] Charques (1938), 32–43, preceded by an article of the same title by the producer-dramatist-critic St John G. Ervine (1883–1971) written from an historical perspective. The book is divided into four parts with twenty contributions. Among other contributors were: commentator Alistair Cooke, actors Reginald Denham and William Devlin, producers Theodore Komisarjevsky and Norman Marshall, Professor Allardyce Nicoll, painter-designer Rex Whistler.

[2] The English actress Dame Peggy Ashcroft (1907–) was performing at the Queen's Theatre 1937–8 (see n. 4 below). She is also referred to in *Footnotes*, 149, 246.

Parrot criticism is predominant; hardly anyone I knew saw *Musical Chairs*[3] without mentioning the word 'Tchehov'.[4]

Cinema audiences are on the whole quicker and more alive to the performance. There is no social kudos in merely going to a cinema, so, with the exception of couples who go there to cuddle, most people go to a cinema to look at the picture. The cinema audience is sentimental and false in its moral values but, trained to Hollywood, it does at least appreciate slickness and efficiency. The lumbering dialogue of many West End comedies would bore it stiff. Whereas I remember going to a comedy in a theatre, *The Composite Man*,[5] where the audience clapped and laughed at jokes which had been seen coming for a dozen lines beforehand and were very flat jokes when they came.

Now who are these people who go to plays and are so easily pleased or so ready to pretend they are pleased? The first point is that, whoever they are, there are not enough of them. Most commercial plays are commercial failures. Any one play, however, will always have its supporters, who support it from force of habit, who judge the pudding by a single stray plum in it and declare accordingly that the whole is witty or daring or profound. These are the habitual theatregoers. When a play is a success it draws many other people who do not go to the theatre habitually. Successful plays usually offer the same values as 'flop' plays; the difference is one of degree. The plums are the same but there are more plums to the pudding. Thus *French Without Tears*[6] is in

[3] *Musical Chairs* (1932), a popular play in three acts by Ronald Mackenzie (1903–32), produced by Theodore Komisarjevsky (1882–1954), with Sir John Gielgud (1904–   ) as Joseph Schindler, at the Criterion Theatre, April 1932. For review, see Derek Verschoyle in the *Spectator*, 148: 5415 (9 April 1932), 506. The play is referred to in *Footnotes*, 144, 247, 256 (illus. opp.), 259.

[4] Among many revivals of Anton Pavlovich Chekhov's (1860–1904) plays in the 1930s, Michel Saint-Denis' production of *The Three Sisters* for Gielgud's repertory season at the Queen's, January 1938, with Peggy Ashcroft as Irina (see n. 2 above), may be mentioned. Chekhov is referred to frequently in *Footnotes*, 59, 83, 137, 145, 199–200, 204, 243, 247, 259, 307, 312.

[5] *The Composite Man* (1935), a comedy about a 'ghost' artist (composer-painter-novelist) by Ronald Jeans (1887–1973), writer of revues, founder of the Liverpool Repertory Theatre 1911, co-founder 1938 with J. B. Priestley (see n. 21 below) of the London Mask Theatre Company. For review of this play produced by the Birmingham Repertory Company, Oct. 1935, see the *Times*, 14 Oct. 1935, 12b.

[6] *French Without Tears* (1936), a successful light comedy by Sir Terence Rattigan (1911–77), founded on a sketch by G. M. Benda; film 1939; musical version 1960 unsuccessful. Produced at The Criterion, Nov. 1936 (running for one year), with

no way a different kind of play from the majority of modern light comedies. It is merely a very good one of its kind. The plums come thick and fast. And light comedy is now so predominant that any other type of play seems novel beyond its merits. Witness the sensation caused by *Mourning Becomes Electra*,[7] a fine play but not in the front rank of high tragedy, and the success of *Jane Eyre*, a bad play whose only asset was a heroine, Miss Curigwen Lewis, who looked the part exactly.[8]

But the theatregoers proper—still we have not decided who they are. I have no statistician to help me, but I would say that they can be roughly divided into two groups. There is a smaller group of pseudo-critics who have canalized in the theatre the universal human passion for being 'in the know' and picking winners. They know all about the actors and are able to criticize acting and production, the sets or the lighting—everything except the essentials of the play itself. The larger group consists of people—mainly women—who use the theatre as an uncritical escape from their daily lives. Suburb-dwellers, spinsters, school-teachers, women secretaries, proprietresses of tea-shops, all these, whether bored with jobs or idleness, go to the theatre for their regular dream-hour off. The same instinct leads them there which makes many hospital nurses spend all their savings on cosmetics, cigarettes and expensive underclothes.

performers Kay Hammond (1909–80), Rex Harrison (1908–), Jessica Tandy (1909–). For review see the *New Statesman and Nation*, NS, 12: 299 (14 Nov. 1936), 772. The play is referred to in *Footnotes*, 213, 225.

[7] *Mourning Becomes Electra* (1931), trilogy on a tragic New England family at the close of the American Civil War, by American playwright Eugene G. O'Neill (1888–1953); based on Aeschylus' trilogy the *Oresteia* ; film 1947. O'Neill won the Noble Prize for literature 1936. London production (107 performances) at Westminster Theatre, Nov. 1937, by Michael Macowan (1906–80), with Beatrix Lehmann (1903–79) as Electra ('Vinny') and Robert Harris (1900–) as Orestes ('Orin'). For review see Desmond MacCarthy, 'A Tremendous Play and Great Acting', the *New Statesman and Nation*, NS, 14: 353 (27 Nov. 1937), 875–7. The play is referred to in *Footnotes*, 68, 149, 218.

[8] *Jane Eyre* (1936), 'a drama of passion in three acts' by Mrs Helen B. Jerome (1883–?), popular adaptation of Charlotte Brontë's novel (1847). Mrs Jerome had also adapted Jane Austen's *Pride and Prejudice* (1813) as 'a sentimental comedy' (1935). For review of *Jane Eyre* produced at the Queen's Theatre, Oct. 1936, see the *New Statesman and Nation*, NS, 12: 297 (31 Oct. 1937), 671–2.

The Welsh actress Miss Curigwen Lewis (*fl.* 1910–40) played numerous parts at the Birmingham Repertory Theatre 1933–6 and the Malvern Festivals 1934–6, incl. *The Moon in the Yellow River* and *1066—and All That* (see nn. 11 and 12 below); she first appeared in London at the Queen's in the title role of *Jane Eyre*, 13 Oct. 1936.

But why, the objector will say, why grudge these people their harmless pleasure? The answer is that it is not harmless and it is not even much of a pleasure. It is a self-indulgent habit like chain-smoking. The chain-smoker, besides hardly tasting his cigarette, loses his taste for other things. And a regular audience which has lost its sense of taste prevents the theatre catering for the wider public who still have a sense of taste but who cannot be counted on to come to the theatre weekly.

The relationship between the theatre and the audience is circular. Bad drama means the wrong sort of audience and the wrong sort of audience means bad drama. There are two ways therefore of improving things. Either educate the theatregoers so that they will demand better theatre, or produce better theatre which will educate the theatregoers (and capture the more educated non-theatregoers). The former course is the more helpful but the less likely to be immediately achieved, as it will depend in its turn on some sweeping change in society. The latter course can at the moment only be pursued in holes and corners. The official theatre is a money-racket, so that any systematic attempt forcibly to intrude better plays into it would fail automatically. Instead, we have all the different types of experimental theatre, whose audiences do not pay so well and are quite differently composed from the audiences of the West End.

Half-way between the West End and the experiments in holes and corners come the provincial repertories and their audiences. The rep. audiences are more intelligent than the West End audiences but they are not intelligent enough. I am not requiring any unusual standard of intelligence in the audience. I merely mean that the higher strata of ordinary intelligence are not properly represented among habitual repertory-goers. The people who regularly go to the Birmingham Repertory Theatre[9] are far from being the most intelligent people in Birmingham. Their superiors

[9] The Birmingham Repertory Theatre—est. 1913 by Sir Barry Jackson (1879–1961) who brought its best productions to London and pioneered 'Shakespeare in modern dress', eventually giving the theatre to the City of Birmingham 1935—was the training ground for such actors as Sir Cedric Hardwicke (1893–1964), Sir Ralph Richardson (1902–83), Dame Peggy Ashcroft (see n. 2 above), Sir Laurence Olivier (1907–), D. Paul Scofield (1922–), Albert Finney (1937–). See Bache Matthews, *The History of the Birmingham Repertory Theatre* (1924); T. C. Kemp, *The Playhouse and the Man* (1943). The Birmingham Repertory is referred to in *Footnotes*, 115, 141, 267–8, 272.

in intelligence could fill the theatre for a week but they do not bother to go to it. People connected with the Birmingham Repertory Theatre sometimes complain that it is not patronized by the University—either by the staff or the students.[10] But if they want a public which shall include members of the University, they must not take as their criterion a public whom they, like the BBC or the Communist Party, assume to be stupid and vulgar. In its earlier days the Birmingham Repertory put on many plays which were vital and intelligent. Financial loss led to their change of policy. They now bank on pieces like *1066 and All That*[11] and occasionally slip in a good play like *The Moon in the Yellow River*[12] which need not be on long enough to lose them their regular clients.

The hole-and-corner organizations, far from trying to compromise with the stupider public, are content to play to a clique. Such cliques are often serious-minded and intelligent but they are also often uncritical—too willing to believe that they are getting something which is art. Similarly, when a dramatic group is closely connected with politics, a good slogan tends to be good enough to carry a play. I went to a Left Theatre Revue where the complacent reverence of the audience was painful to contemplate.[13] Shoddy writing, production, and acting were

[10] Birmingham University did not become formally associated with the Repertory Theatre until 1962, a year after Sir Barry's death.

[11] *1066—and All That* (1935), musical comedy by the English librettist Reginald Arkell (1882–1959), based on the book of burlesqued history (1930) by Walter C. Sellar and Robert J. Yeatman, music by the English composer-conductor Alfred Reynolds (1884–1969). Produced at the Strand Theatre, April 1935; revived at the Cambridge Theatre, March, 1937. For reviews, see *The Times*, 26 April 1935, 14b; 29 March 1937, 8e.

[12] *The Moon in the Yellow River* (1931), second play by Denis W. Johnston (1901–84), widely popular for diagnosis of the mid-1920s' mood in Ireland. For review of Malvern Festival production, July 1934, see Graham Greene, *Spectator*, 153: 5536 (3 Aug. 1934), 161. For review of Westminster Theatre (London) production, Sept. 1934, see Desmond MacCarthy, 'John Bull's Other Island Up to Date', *New Statesman and Nation*, NS, 14: 344 (25 Sept. 1937), 443–4.

Irish writer, director of the Dublin Gate Theatre 1931–6, Johnston was also a broadcaster and professor of English. He wrote a remarkable autobiography *Nine Rivers from Jordan* (1953) about his experiences as a BBC war correspondent in the Second World War.

[13] The Left Book Club founded 1936 by Victor Gollancz (see n. 14) established a Theatre Guild movement, run by John Allen, organized from Unity Theatre, St Pancras, from May 1937, to put on leftist plays, reviews, skits: John Lewis, *The Left Book Club* (1970), 46–8. In 1937–8 there were 250 branches of the Guild.

uproariously applauded, under the aegis of Mr Victor Gollancz[14] and the shadow of Spain.[15]

Reverence is, however, a good thing. The European theatre began in religion. Even supposing we acquire a new communal religion, we shall never reinstate the drama as something essentially religious, for the simple reason that many centuries have irrefutably proved that the theatre *can* be secular. But a compromise is possible between the view that drama is an act of worship and the view that it is mere entertainment. Like poetry, the drama ought to be connected with life—ought, in Matthew Arnold's phrase, to be a 'criticism' of life. This being granted, it seems desirable that the drama should present or 'criticize' either that kind of life which is comprehensible to the majority of the public or that kind of life which is comprehensible to the most intelligent or most vital sections of the public. It is obvious that no branch of modern English drama—West End, repertory, or hole-and-corner—does either of these things.

The best example of the wedding of a high and subtle drama to an intelligent and fairly representative audience is offered by Athens in the fifth century BC. There are three points which strongly distinguish it from our modern drama. (1) Athens was a small city state; the Athenian theatre held twenty thousand people; it was therefore possible for a very large section of the whole population to see any one play simultaneously. (2) Plays were produced in Athens on only two occasions during the year.[16] They were therefore an event to be waited for, like the Calcutta Cup match.[17] There was no chance of developing the vice of chain-smoking. (3) In the case of tragedy, the audience shared with the dramatist a common ground—the traditional mythology—on which serious and intricate plays could be built

[14] Sir Victor Gollancz (1893–1967), writer-publisher (publisher of MacNeice's first book *Blind Fireworks*, 1929), sponsor of political and humane causes, the Left Book Club in the 1930s and the Association for World Peace 1951; writer of autobiographies, anthologies, musical companions.

[15] Spanish Civil War 1936–9.

[16] i.e. at Dionysiac festivals—tragedy in the spring (March–April), comedy in the winter (Jan.–Feb.): see Sir Arthur Pickard-Cambridge, *The Dramatic Festivals of Athens*, 2nd edn (1968).

[17] 'Calcutta Cup, Rugby Union perpetual challenge trophy at issue since 1878 in the annual match between England and Scotland': *The Oxford Companion to Sports & Games*, ed. John Arlott (1975), 154.

up with an economy of effort and with comparatively little fear of obscurity.

In the British Isles to-day we lack a mythology and, further, we lack a community creed. There are vast numbers of people with different creeds or no creeds, with different attitudes to life. It is impossible for a modern English dramatist to write for the whole British Isles in the same way as Aeschylus wrote for Athens or even for Greece. In the Soviet Union we are told that a greater unity of outlook has led all over the country to an efflorescence of popular drama.[18] Some great change in the constitution of our society might lead to a similar efflorescence. In the meantime it looks as if we shall have to make shift with something like cliques—writing for small specialized sections of the public but preferably in such a way that wider and wider sections can understand us and appreciate us if they want to. Thus Stephen Spender's play, *The Trial of a Judge*, [19] produced by the Group Theatre in London for an audience of highbrows and a small section of Left Wing political sympathizers, must, in my opinion, have a far larger audience waiting for it through the length and breadth of the British Isles.

The dramatist ought to be in touch with contemporary life (which, strictly speaking, is the only life) and ought to reflect that life to the people who are acquainted with it. But everyone, the objector will say, is acquainted with it. This is of course true, but there are degrees of being alive as there are degrees of truth and reality. The lunatic is not properly alive (nor, some would add, is the capitalist). The West End theatre, for the most part, presents the half-alive to the half-alive. Or, what is worse, it presents the half-alive to the three-quarters-alive and persuades the latter that they are seeing the portrait of their betters. In any case it presents something to people who are not in a position to criticize it. It presents the drawing-rooms of an obsolete class—the rich and leisured—to people who live in semi-detached villas and have never even met a *divorcée*.

[18] For Soviet Russian theatre, see Sylvia Saunders, 'The Theatre in the Dictator Countries: (1) Russia', *Footnotes*, 167–75; for the Moscow Art Theatre, see *Footnotes*, 59, 83, 94, 169, 171–2.

[19] A tragic verse-drama first produced by Rupert Doone at the Group Theatre on a bare stage with minimal décor by the English painter-designer John Piper (1903–), March 1938. For theatre review see *The Times*, 19 March 1938, 10d.

But what, the objector will say, about the plays of Shakespeare? Shakespeare who treated the groundlings to kings and dukes. But Shakespeare's plays are not naturalistic studies of kingship and dukeship. Kings and dukes were the occasion for the working-out of universal human feelings and problems. This is not the case in the West End drawing-room play. *Lear* and *Othello* flowed from contemporary life at its richest.[20] Whereas Noël Coward's surface limitations of subject-matter are actual limitations.

Many modern plays flow at least from contemporary ideas; the greatest example is Shaw. But man cannot live by ideas alone. The discussion play tends to have the same vicious appeal as books of popular science—How to be In the Know for Seven-and-Six. Hence the popularity of inferior plays like *I Have Been Here Before*.[21] Many people of very different kinds liked this slow and sticky play because they were captivated by the theory about time on which it is strung—'such an interesting idea'. To contain an idea is certainly an asset for a play, but rather than go to a bad play which contains a good idea I would subscribe to the star system and go to a bad play like *Autumn* in order to see a good actress like Miss Flora Robson.[22]

Yet I agree that the star system is vicious. The audience will not be able to appreciate drama properly as long as they go to the theatre with their minds sizzling with the irrelevancies of gossip columns. The poet is hampered by his public's misconception of him as someone essentially different from themselves. So it is with the playwright and the actor. The public ought to think of them

[20] Shakespeare and the Old Vic are cited frequently in *Footnotes*, with many plates of productions; *Lear* is referred to at 28, 261, and *Othello* at 83, 261.

[21] *I Have Been Here Before* (1937), psychological play by John Boynton Priestley (1894–1984), the English dramatist, novelist, critic, based on the time theories of the Irish aeronautic expert John William Dunne (1875–1949), produced at the Royalty Theatre, Sept. 1937, by Sir Lewis T. Casson (1875–1969) who also played the Dunne *persona* Dr Gortler. For review see Desmond MacCarthy, *New Statesman and Nation*, NS, 14: 345 (2 Oct. 1937), 486–7. Priestley is referred to in *Footnotes*, 144, 213.

[22] *Autumn* (1937), play by Margaret Moore Kennedy (1896–1971; already known for her novel and play *The Constant Nymph*, 1924 and 1926) and Gregory Ratoff (1893–1960; the Russian actor and film director), adapted from the Russian *Autumn Violins* (1915) by Ilya D. Surguchev (1881–1956), produced at St Martin's Theatre, Oct. 1937, with Flora Robson as Lady Catherine Brooke. For review see *The Times*, 16 Oct. 1937, 10d. The English actress Dame Flora Robson (1902–84) is referred to in *Footnotes*, 45, 148, 273.

as their spokesmen and be as willing to criticize them as they are to criticize their democratically elected Members of Parliament.

The relationship between audience and actors would need a delicate and qualified analysis. In a sense the audience co-operate in the making of a play. But only in a sense. I like Mr Christopher Isherwood's image of the theatre as a kind of box in which audience and actors are imprisoned together and from which they can only be released by a key which is the dénouement of the play.[23] But it is the actors who are active in discovering the key. This job has been entrusted to them by the audience. It is a mistake to run to extremes and claim that the audience themselves are also actors. The audience undergo communion with the actors, and for that reason certain devices are desirable which will prevent them feeling too cut off from the stage. Thus in the Ancient Greek theatre the orchestra in front of the stage and in the centre of the semicircle of seats acted as a bridge between the stage-world and the real world in the same way as did the chorus which performed in the orchestra. The apron-stage of the Elizabethans allowed soliloquies and asides to get home to the audience instead of being wasted in the air yards away behind the proscenium arch. And I have heard actors complain of the Shakespeare Memorial Theatre at Stratford that they cannot ever feel at home on its stage because there is such a gap between the footlights and the front row of the stalls.[24]

But the drama, as well as being communion, is also essentially spectacle, and the spectator must be able to watch it undisturbed. For this reason I disapprove on the whole of what used to be the favourite trick of the Festival Theatre at Cambridge—the regular

[23] Probable allusion to a Group Theatre paper or programme note; untraced.

[24] The new Shakespeare Memorial Theatre, Stratford-upon-Avon (the old Memorial Theatre burned to the ground 1926), designed with proscenium arch but without apron-stage by Elisabeth Scott (1898–1972), opened 23 April 1932. See Ruth Ellis, *The Shakespeare Memorial Theatre* (1948). 'The fundamental weakness in the design of the [Shakespeare] Memorial Theatre is the gulf between stage and auditorium. This would be a serious enough defect in any theatre, but is doubly so in a theatre built for the plays of Shakespeare which were written for a platform stage with no proscenium arch and no barrier of any sort between actor and audience': Norman Marshall, *The Other Theatre* (1947), 176. The theatre is referred to in *Footnotes*, 103, 272, and its stage productions frequently illustrated therein.

entry of actors through the auditorium.[25] I cannot watch the stage with proper attention if I am always expecting someone to come hooting and blustering past me from an entrance at the back of the house. In the early days of the Soviet Union producers went to extremes in this direction. Vakhtangov and Meyerhold[26] scrapped all stage illusionism and told their actors to dress and make up on the stage, to mingle with and talk to the audience, while the audience in their turn were encouraged to join in the performance. This was a healthy stage of experimentation, but it could not last. Mob self-expression is out of place in the theatre, and modern Russian audiences are, I am told, lapsing once again to the role of the idle looker-on—idle, that is, in body but not in mind.

Lastly, what sort of plays do I want and what sort of audience do I want for them? I think that plays should be written about the important problems which fill the lives of nations and individuals. I would not, of course, exclude the purely frivolous light entertainment, but I do not think that it should be allowed either to dominate the theatre or to pretend that it is something which it is not. Plays on specialized subjects should be written by people who know them from the inside; witness the early plays of Sean O'Casey.[27] Plays on generalized subjects should probably be written by poets, using probably some non-naturalistic technique.

[25] The Festival Theatre, Cambridge, was founded 1926 by Terence Gray (1895–) who spread advanced ideas on lighting and stagecraft begun 1903 by E. H. Gordon Craig (1872–1966). Gray abandoned the project 1933 when the theatre was bought by a commercial management and reconverted to conventional style. See Marshall, op. cit., chap. 5 ('The Festival Theatre, Cambridge'). MacNeice in this para. insists upon æsthetic distance in drama, against experimental violations of stage illusion or spectacle.

[26] Eugene V. Vakhtangov (1883–1922) and Vsevolod E. Meyerhold (1874–?1943), the Soviet Russian actor-producers, who worked under Konstantin E. Stanislavsky (1863–1938) at the Moscow Art theatre; then, influenced by his anti-histrionic naturalism (which became the Stanislavsky Method), founded their own original theatres for experimental and revolutionary plays: the Vakhtangov Theatre continued under his widow, but the Theatre of Meyerhold was closed 1938. Meyerhold is referred to in *Footnotes*, 94–5, 171.

[27] The early works of the popular Irish dramatist Sean O'Casey (1880–1964) were the realist plays *The Shadow of a Gunman* (1923), *Juno and the Paycock* (1924), *The Plough and the Stars* (1926), produced at the Abbey Theatre. O'Casey's later plays, written in exile in England after 1926, used expressionist and symbolist techniques and did not find as wide an acceptance as his early plays written for the Abbey.

Juno was produced at the Haymarket, London, July–Aug. 1937, with experienced Abbey players. *Plough* had been made into an American film 1937, adapted

That people can bear to watch poetic plays is shown by the success of *Murder in the Cathedral*.[28] Few, I suspect, would have gone to this just because it was poetry, but many went because they were believers and they liked their beliefs to be more memorably treated than they could be in ordinary prose dialogue. For poetry is in at least one respect more true to human nature than prose, because human nature is itself essentially rhythmical. Belief and rhythm—the two things least evident in our modern theatre—are the two things most to be desired in it.

My hopes for the theatre would rest mainly on a change in the structure of society but failing that, on the spread of secondary education. (Thus in America there is a much greater percentage of the population fit to go to the theatre.) On the other hand, I would demand a fairer distribution of theatres throughout Greater London and the provinces, the existence of regular touring companies, lower prices everywhere, less snobbery. I would shake up the actors (I cannot see why an ignorant and stupid person should be the most qualified to act) and I would, I think, scrap the professional dramatic critics. Let each paper have a list (periodically changed) of some fifty intelligent theatregoers who can take it in turns to write the notices of plays. I would have no brow-beating of the public by superior persons. The drama is one of the highest of art-forms but it is one which depends for its success on the co-operation of many. And it cannot be successful without an audience. Let the drama encourage the audience and perhaps the audience will encourage the drama.[29]

by Dudley Nichols (1895–1960), directed by John Ford (1895–1973) for RKO Radio. For review of the Haymarket *Juno* see the *New Statesman and Nation*, NS, 14: 338 (14 Aug. 1937), 252. For review of the *Plough* film see Mark Van Doren, 'Considering the Source', *Nation*, 144: 7 (13 Feb. 1937), 194. Sean O'Casey is referred to in *Footnotes*, 71, 143, 272, 310.

[28] Successful verse drama by T. S. Eliot initiating a revival of poetic drama—based on the martyrdom of Saint Thomas à Becket and first produced in the Chapter House of Canterbury Cathedral by E. Martin Browne (1900–80), June 1935, with Robert W. Speaight (1904–76) as the Archbishop. After a run at the Mercury Theatre, the play was seen at the Duchess, the Old Vic, and on tour in England. For reviews, see John Pudney in the *New Stateman and Nation*, NS, 9: 226 (22 June 1935), 927–8; *The Times*, 17 June 1935, 10e; etc.

[29] The mixed reception accorded the Group Theatre productions of MacNeice's plays the *Agamemnon* and *Out of the Picture* 1936–7 may account for the strong views expressed in this final para. See theatre notices of these productions (Nov. 1936; Dec. 1937) in *The Times*, 2 Nov. 1936, 12c; 6 Dec. 1937, 18c; cf. Michael J. Sidnell, *Dances of Death* (1984), chap. 10 ('Louis MacNeice and the Group Theatre').

A Statement[1]

I have been asked to commit myself about poetry.[2] I have committed myself already so much *in* poetry that this seems almost superfluous. I think that the poet is only an extension—or, if you prefer it, a concentration—of the ordinary man.[3] The content of poetry comes out of life. Half the battle is the selection of material. The poet is both critic and entertainer.[4] He should select subjects therefore which (*a*) he is in a position to criticize, and (*b*) other people are likely to find interesting. The poet at the moment will tend to be moralist rather than æsthete. But his morality must be honest; he must not merely retail other people's dogma. The world no doubt needs propaganda, but propaganda (unless you use the term, as many do, very loosely indeed) is not the poet's job. He is not the loud-speaker of society, but something much more like its still, small voice. At his highest he can be its conscience, its critical faculty, its grievous instinct. He will not serve his world by wearing blinkers. The world to-day consists of specialists and intransigents. The poet, by contrast, should be synoptic and elastic in his sympathies. It is quite possible therefore that at some period his duty as a poet may conflict with his duty as a man. In that case he can stop writing, but he must not degrade his poetry even in the service of a good cause; for bad poetry won't serve it much anyway. It is still, however, possible to write honestly without feeling that the time for honesty is past.

[1] *New Verse*, nos 31–2 (Double Number on 'Commitments', Autumn 1938), 7.

[2] Public declarations of commitment became fashionable among leftist writers and artists in the later 1930s, under pressure of international political crises. MacNeice's objection was echoed by Spender, 'Left Wing Orthodoxy', *New Verse*, 31–2 (Autumn 1938), 15: 'it is essential to judge work by its achievement and not by the opinions of the writer.'

[3] Stated by MacNeice in *Modern Poetry* (Nov. 1938), 1, 197–8.

[4] Cf. ibid., 197: 'I consider that the poet is a blend of the entertainer and the critic or informer; he is not a legislator . . . nor yet a prophet'.

The Oxford Book of Light Verse[1]

The Oxford Book of Light Verse. Chosen by W. H. Auden. (Oxford 8*s*. 6*d*.)

This is one of the most delightful anthologies in English. It would be merely academic to cavil at Mr Auden's interpretation of the term 'Light Verse'. His range is very (and oh how refreshingly) wide and he gathers together pieces as different as Pope's 'Epistle' to Dr Arbuthnot and the doggerel broadsheet on the death of King Edward VII. Yet while the selection emphasizes the manysidedness of poetry (which cannot be emphasized too often) the great majority of these pieces *have* something in common which distinguishes them from Matthew Arnold's poetry of 'high seriousness'.[2] This is not to say that many of these poems are not serious (Aristophanes after all was a serious writer), but their seriousness is not, like that of Milton or Shelley, regularly on the high horse; it is less rarefied, less pure, less self-conscious, and therefore more congenial to the ordinary individual or to any one individual's ordinary moments.

Mr Auden in his most pregnant introduction deplores the narrowing of the term 'light verse' to *vers de société*, etc., and in his selection has included comparatively little of this genre. This need not be regretted. There are many anthologies devoted to such poems which tend in any case to pall when read in bulk. Light verse of this kind is as near as possible light all through, being usually the outcome of superficial armchair cerebration and not representative of a genuine feeling or a naturally felt attitude to life. Popular verse on the other hand flows from a genuine feeling or a genuine attitude; this is the case with nursery rhymes, broadsheets, folk ballads; one may note in passing that the *form* of these is usually as effective for its purpose as the form of the professionals.

[1] *Listener* 20: 514 (17 Nov. 1938), 1079; unsigned review. The anthology is dedicated to Professor E. R. Dodds of Oxford, mutual friend and mentor of MacNeice and Auden from Birmingham days.

[2] Arnold's criterion for poetry in 'The Study of Poetry' (1880).

Many of the finest works of art are a conscious or (more often) unconscious compromise, the outcome of a dialectic of opposites—heroics tinctured with the grain of salt or satire leavened with sympathy; witness the Odes of Horace or Villon's Testament. Thus Mr Auden's book contains many modern ballads which on the face of it are lighthearted—'Frankie and Johnny', 'John Henry', 'Johnny, I hardly knew ye', and even 'John Peel'—but which imply a genuine, and sometimes a tragic, 'criticism of life'.[3] In the same way Lear's nonsense poetry, represented here by 'The Jumblies', 'The Owl and the Pussy-cat', 'My Uncle Arly', and 'How pleasant to know Mr Lear!' can be regarded as the last outcrop of the Romantic Revival with its nostalgia, its self-pity, its anarchism, its grandiosity.

Chaucer, Skelton, *Hudibras*, Dryden, Pope, Swift, Burns and Byron are all naturally represented here. Blake's squibs about Reynolds and Rubens are in place though his 'Auguries of Innocence' is one of the few poems which should not perhaps have been included. It is good to meet Appleton House[4] again, but the poetry of conceits might also have been represented by Thomas Carew.[5] There are poems by A. E. Housman, Bridges, Kipling, Chesterton, Belloc, D. H. Lawrence and John Betjeman, but it is a pity that Mr Auden did not put in any of his own; he is one of the few living poets whose poetry can walk in the street without falling flat on its face.

As a comment on poetry or the nature of poetry or the function or anything-else-you-like of poetry, this book is worth a hundred laboured volumes of literary criticism. And there must be few readers who will not discover here many superb poems previously unknown to them—poems to fit nearly all their everyday moods, their sceptical, sentimental, venomous, epicurean, nursery, music-hall, tongue-in-the-cheek or devil-may-care moments.

[3] Arnold's definition of poetry in ibid.

[4] Andrew Marvell's long poem 'Upon Appleton House' (1650s).

[5] Thomas Carew (?1594–1640), writer of sensual and charming love poetry after the courtly manner of Ben Jonson.

On the Frontier. By W. H. Auden and Christopher Isherwood[1]

On the Frontier. By W. H. Auden and Christopher Isherwood. At the Arts Theatre, Cambridge

Being one of those many people who consider Mr Auden and Mr Isherwood to be respectively the best of the younger writers in verse and in prose, I am always expecting great things from their collaboration in drama. Thus *The Ascent of F6*, though very open to criticism, seemed to me to be one of the most exciting and moving plays I had seen for a long time—a most significant theme handled with great ingenuity.[2] After this *On the Frontier* was disappointing. The theme again is vastly important, but the treatment is rather facile and seems to fall between two stools as if the authors could not decide whether they were writing a straight play or a crooked one. Compared with *The Ascent of F6*, there is less sparkle, less poetry, less thought and even more embarrassment. The mystical love scenes of Eric and Anna made one long for a sack to put one's head in.[3]

It would be foolish, however, to regard *On the Frontier* as a step downhill. Its theme—the horrible complex of international rivalries, crooked big business, Fascism, self-deceiving heroics and hysterical publicity—is a theme which cries to be dramatized; the problem is merely how to do it. Mr Auden and Mr Isherwood have tried to do it by a series of melodramatic cartoons. Anyone can see the point of these; the sentiments are admirable; occasionally we are excited or amused or moved but the play as a whole lacks cohesion; it does not hit us like a wedge but like a number

[1] *Spectator* 161: 5760 (18 Nov. 1938), 858. For the Cambridge production, partly backed by J. Maynard Keynes (subsequently it moved to the Globe Theatre, London), see Sidnell, *Dances of Death*, chap. 12.

[2] *The Ascent of F6* (1936) was produced by Rupert Doone and the Group Theatre with music by Benjamin Britten at the Mercury Theatre, 26 Feb. 1937; then at the Arts Theatre, Cambridge, 22 April 1937: see Sidnell, chap. 9.

[3] Ends of scenes I. ii, II. i, III. iii. In the finale (III. iii) Anna says to Eric, 'The place of love, the good place. / O hold me in your arms. / The darkness closes in.' [*The lights fade slowly. Background of music.*] Their earnest prayerful verse recitations were embarrassing in simplistic sentimentality.

of escaped posters and photographs blown by the wind in one's face. War cartoons have their function but I feel that Mr Auden with his poet's knack of palpable generalization and Mr Isherwood with his novelist's knack of psychological observation will in the future treat the same theme again more solidly and therefore more memorably.

An appreciative audience came to the first performance on November 14th at the Arts Theatre, Cambridge, an excellent place for the first production of that rare thing, a serious play. This audience, in spite of the above criticisms, certainly got its money's worth. (A not entirely successful play on a really important subject is, after all, a great deal more worth while than an alpha-plus domestic triangle among the sherry glasses.) The production, like most Group Theatre productions, was alive and intelligent; Mr Doone, a producer of the intuitive school, improves from play to play.[4] The two main characters, Valerian, the business Machiavelli, and the Dictator, were excellently played by Mr Wyndham Goldie and Mr Ernest Milton.[5] Mr Milton's Dictator, alternately abstracted and hysterical, was only too convincing. Mr Goldie's Magnate, perfect in his cynical virility, was the only one of those characters whom I should like to meet in the life. This is the most interesting part and I feel that the authors would have done well to centralize the whole play around him more than they have done. The women's parts were unsympathetic (a play ought to have one woman one can like in it). Miss Everley Gregg was impressive as the fanatical spinster[6] on whom the authors (compare Mr Auden's 'Ballad of Miss Gee')[7] sadistically visit the sins of her society, striking her with plague

[4] 'Rupert Doone' (pseud. of Ernest Reginald Woodfield, 1903–66): see Sidnell, 308–10.

[5] Frank Wyndham Goldie (1897–1957), the English actor, lead in the Liverpool Repertory Company 1927–34, leading West End actor from 1934, taking on film and later television work. Ernest Milton (1890–1974), popular Anglo-American actor in theatre 1912–67, entered films 1935.

[6] Everley Gregg (Mrs Winston Walker, 1903–59), the English actress, played Martha Thorvald, described in 'Notes on the Characters' as 'Violently repressed, fanatical. Wears glasses. . . . Beneath her fanaticism, she is an educated, intelligent woman.'

[7] 'Miss Gee' (1937), a grotesque satirical ballad about a spinster who dies of cancer 'caused' by sexual repression: *The English Auden*, 214–16. For Auden's 1930s doctrine 'that physical disease is always symptomatic of a psychological cause'—taken from John W. Layard (1891–1974), disciple of the American 'healer' Homer Lane (1876–1925)—see Spears, *The Poetry of W. H. Auden*, 8.

in front of the portrait of her leader.[8] Miss Lydia Lopokova as the beloved on the astral plane had no opportunity to show her remarkable vivacity.[9]

Between the scenes of the play there are musical interludes, the music by Mr Benjamin Britten[10] being both impressive and appropriate. And the sets designed by Mr Robert Medley were excellent.[11] This brilliant designer here had an opportunity to be more elaborate than usual and he took it with both hands. Valerian's study was well worthy of Mr Goldie while the difficult problem of the Ostnia-Westland room (i.e., the splitting of the stage into two family parlours in adjoining countries) was solved so as to please and even to convince.[12]

Both Mr Auden and Mr Isherwood believe in literature as a criticism of life. In the narrower sense this play is excellent criticism, for it is difficult not to agree with the moral or the morals of it. But a play cannot live by morals alone and I look forward to the day when this dangerously talented partnership, so prolific of brilliant silhouettes, gives birth to a more plastic set of characters.[13]

The *Antigone* of Sophocles[1]

'The *Antigone* of Sophocles.' At the Arts Theatre, Cambridge.

The *Antigone* is a magnificent play and one of the few Greek plays in which the conflict is immediately intelligible to a modern audience (even though we do not attach much importance to that

[8] End of III. i.

[9] Lydia Lopokova (1891–1981), born in St Petersburg, a ballet dancer with Diaghilev's Ballet Russe, actress 1914–39, wife of Baron Maynard Keynes (married 1925), played Anna Vrodny.

[10] English composer (later Lord Britten of Aldeburgh, 1913–76), to whom the play was dedicated.

[11] For Robert Medley (1905–), see his *Drawn from the Life: a memoir* (1983); also Sidnell, *passim*.

[12] i.e., the frontier was represented by a divided stage: I. ii, II. i, III. i.

[13] This play was the last collaboration in drama between Auden and Isherwood; their travel book *Journey to a War* was published March 1939.

[1] *Spectator* 162: 5776 (10 March 1939), 404. *The Times* reviewer—in 'Sophocles at Cambridge: The "Antigone" in Greek', 6 March 1939, 12d—regarded this production as timely in view of 'the troubles of the moment' and so as proof of the play's 'universal application'.

formal burial of the dead which is the particular occasion of this conflict). The opposition between the law of the State (in this case a dictatorship) and natural morality is given two solid representatives in Creon and Antigone. My chief criticism of the present Cambridge production is that Creon was not solid enough. It can be argued that Creon ought to be played as a weak, neurotic character who tries to cover up his weakness with obstinacy and ruthlessness. But I think this would be reading modern conceptions into it. One should beware of treating a Greek play—with the exception of some of Euripides—psychologically. The *muthos*, as Aristotle said, is the important thing.[2] Creon and Antigone embody two opposing principles and Creon must at all costs be the appropriate vehicle of his principle.

On the whole the Provost of King's[3] gave the *muthos* a good deal in this production, the whole action progressing convincingly from beginning to end, though it seems a pity that there had to be an interval. For those who have little Greek Mr Patrick Hadley's music[4] (which was less churchy than such music often is) must have bolstered up the dialogue admirably. The chorus, who looked very well in their archaic beards, were nicely managed and it was a good idea not to have them all on the stage all the time. From the front rows of the stalls it is difficult to see the grouping of actors and chorus as a unity—an inevitable snag in the production of any Greek play where the chorus has to be on the stage.[5] I think it was, perhaps, a mistake to dress the three chorus leaders differently from the rest and beardless; surely the leaders should, if anything, look older than the rest.

[2] 'So Plot [*mythos*] is the basic principle, the heart and soul, as it were, of tragedy': Aristotle, *Poetics*, 6, 1450^a 38–9; trans. Else, 28.

[3] J. T. Sheppard: see n. 7 below.

[4] Patrick A. S. Hadley (1899–1973), Professor of Composition at the Royal College of Music from 1925, had returned to Cambridge 1938 as University Lecturer in Music and took charge of Caius Chapel music; he was elected Music Professor 1946. His new music for the choruses of Sophocles' *Antigone* used 'some curious instruments, incl. an organ, and elaborate very dramatic orchestration' (*The Times*, loc. cit.). Heretofore Mendelssohn's music for *Antigone* had been both familiar and successful—Sir Richard Jebb's 3rd edn of *Antigone* (1900, 1962), xli.

[5] Traditionally the Chorus in a Greek amphitheatre used not stage but dancing floor (*orchestra*).

The set was simple and impressive: a palace of Mycenean blocks[6] with a doorway large enough to hold the corpse of Eurydice longways; also an altar with real incense on it. I should, however, have liked a higher rostrum at the entrance to the palace (a desire, perhaps, accounted for by my position in the auditorium). It also seemed a pity that Creon should have to come so far down stage to make his speeches, which on every occasion necessitated a long (and unkingly?) retreat with his back to the audience. Mr Sheppard's[7] most debatable experiment was after the exit of Creon to release Antigone. Here the lights became brighter and we suddenly, for the first and last time, had some eurhythmics and hand-clapping (executed, fortunately, by soldiers and not by the chorus) before the lights dimmed to denote the collapse of this spasmodic upburst of hope. A good idea, which, I think, came off, though it came dangerously near to vulgarity.

Mr Tregoning, who looked the part, gave a fine performance as Antigone.[8] Mr Scott-Malden, as the Watchman,[9] made a success of it as a character part, though perhaps overdoing the zany touch. Mr Goodall, as Haemon,[10] looked, except for his stature, like the Apollo at Olympia.[11] But Mr Millar, as Creon,[12] did seem to me to be wrongly cast or wrongly produced. His features were too feminine and, although he spoke well, he lacked both the voice and the presence, and many of his movements were jerky and ineffectual. It is, admittedly, an extremely difficult part.

The verse was uniformly well spoken, though I cannot approve the retention of the 'older' Greek pronunciation with its plethora

[6] Probably modelled on the Lion Gate or the Treasury of Atreus at Mycenae: see Alan J. B. Wace, *Mycenae* (1949, 1964), illus. 72–4 and 39–46.

[7] Producer: John Tressider Sheppard (1881–1968), later Sir John, MBE, Provost of King's College 1933–54, famous in the 1920s and 30s as lecturer, translator, and producer of plays in Greek: *The Times* obituary, 9 May 1968, 12e.

[8] Peter Norris Tregoning of Trinity (BA 1939) as the heroine Antigone.

[9] F. D. S. Scott-Malden of King's as the Watchman.

[10] R. G. Goodall of King's as Haemon, Creon's son, Antigone's betrothed.

[11] For this Phidias sculpture (original height 3.15 metres) in the Temple of Zeus at Olympia, see Gerhart Rodenwaldt, *Olympia*, photo. Walter Hege (1936), frontispiece, text 39–40, illus. 43–8: 'the divine ideal of a heroic age'.

[12] R. G. Millar of King's as Creon.

of 'ow' sounds.[13] The sung choruses, of course, one could not disentangle. When a Greek play is done in translation this would be objectionable, but in an all-Greek production music is both a relief and an enhancement. Mr Geoffrey Wright's costumes were sometimes good—e.g., the chorus—and sometimes unsatisfactory—notably Creon, who looked like a king from a puppet-show.

Taking it as a whole, however, I consider this a most commendable production and much preferable to that of the potted *Oresteia* in 1933[14]—the last Cambridge Greek play which I saw. It is, of course, a more sympathetic subject. The Hereditary Curse has gone, but Creon is always with us, and we still can find inspiration in the individual conscience at bay.

Original Sin[1]

The Family Reunion, by T. S. Eliot. (New York: Harcourt, Brace and Company. 131 pages. $1.50.)

The Family Reunion seems to me a better play than *Murder in the Cathedral*,[2] better integrated, less of a charade. This time the subordinate characters are real persons, fuller, more differentiated,

[13] On the history of the pronunciation of classical Greek, see W. B. Stanford, *The Sound of Greek* (1967), chap. 6 ('Matters of Pronunciation'), esp. 129 for pronunciation of Greek *ou* (English 'ow' in old fashion; 'oo' in reformed manner); also W. Sidney Allen, *Vox Graeca*, 2nd edn (1974). 'The reformed pronunciation of Greek came into use in English-speaking countries concurrently with that of Latin, and in consequence has long been the normal practice in Scotland and the U.S.A. In England it has become widely adopted since its recommendation by the Classical Association's Committee on the Pronunciation of Greek (1908): but there are still a few schools that have joined the vast majority in employing the reformed pronunciation of Latin, yet continue to pronounce Greek in the antiquated "English" way': *The Teaching of Classics*, issued by the Incorporated Association of Assistant Masters in Secondary Schools, 2nd edn (1961), App. II, 219.

[14] J. T. Sheppard's Cambridge adaptation of Aeschylus' trilogy the *Oresteia* in Feb. 1933 had been abridged to fit into a single performance and took stage liberties shocking to some Greek scholars: *The Times* review, 'Aeschylus at Cambridge: The "Oresteia" ', 15 Feb. 1933, 10b; see also letter of Sir Charles Strachey, *The Times*, 6 March 1933, 8e.

[1] *New Republic* 98: 1274 (3 May 1939), 384–5.

[2] Eliot's liturgical drama, presented at the Canterbury Festival, June 1935: see MacNeice's article 'The Play and the Audience' (1938) above, n. 28.

more sympathetic; and the ideas behind the play are fused into the action and the characters; it is difficult (and this is as it should be) to divorce the theme or the moral from the play itself. It would be an easy play to ridicule—a hag-ridden hero[3] who appears in a vague mess and disappears toward a vague solution—but such ridicule would be misplaced. Aristotle thought that the soul of a play is action.[4] If we interpret action in the narrow or external sense, then according to Aristotle this play is not dramatic. But Mr Eliot has always been more interested in action, and in the correlative suffering, on the spiritual plane. His religious beliefs, as can be seen from such books as his notorious *After Strange Gods*, have opposed him to 'liberalism', to any basically utilitarian doctrine of progress.[5] From one point of view, then, Mr Eliot is a reactionary, but he is at the same time a corrective to the facile optimism of many Leftist writers. We may regret that he seems to put all his money on the religious conscience as distinct from practical morality, but at the same time we must recognize that he asserts certain truths (even if these are the truths of the Unknown God) which are now commonly neglected and whose neglect may in the long run sap the life from our utilitarian ethics:

> . . . the circle of our understanding
> Is a very restricted area.
> Except for a limited number
> Of strictly practical purposes
> We do not know what we are doing.[6]

Though the subject of his play is Original Sin, Mr Eliot has embodied it in characters who on the surface plane also are involved in dramatically interesting relationships to each other (this set of characters in the same situation could in fact have been treated by Chekhov). There is a compromise here between naturalism and mysticism.[7] The definite surface facts—the

[3] Harry, Lord Monchensey.

[4] Cf. MacNeice's comment, 'The *muthos*, as Aristotle said, is the important thing,' in his review 'The *Antigone* of Sophocles' (March 1939) above with n. 2.

[5] *After Strange Gods: A Primer of Modern Heresy* (1934), a conservative, orthodox critique of modern literature, was never republished by Eliot.

[6] *The Family Reunion* (1939), II. iii, p. 128: Eliot's *Collected Plays* (1962), 120.

[7] In his poetic dramas Eliot consciously exploited two planes in the action, realistic and symbolic-religious: see his 'John Marston' (1934) in *Selected Essays*, 229. MacNeice attempted a similar doubleness in his parable plays for radio.

mother's birthday, the family house, the brothers' accidents, the hero's home-coming, the previous death of his wife, the death in the last scene of his mother—may be from Mr Eliot's point of view merely incidental, but they act as girders to the play. Thus the hero, like Orestes,[8] has apparently committed (or thinks he has committed) a murder;[9] this murder is merely incidental to, or at most symptomatic of, a far more basic and less particularized sin which he has to expiate. The Eumenides[10] who haunt him appear at first sight to be subjective phantoms but are discovered, to the hero's own belief, to be forces outside him. His expiation on the face of it seems to consist in leaving his home forever; this is in fact the outward and visible sign of a profound spiritual change. This change being still obscure, Mr Eliot was of course right to stress the outward and visible signs. For this reason the play seems to me more suited to the stage than *Murder in the Cathedral*.

The trouble with *Murder in the Cathedral* was that the essential conflict was between Becket and himself as represented by the Tempters; the murderers[11] merely arrived out of a machine. In *The Family Reunion*, the hero is again struggling with himself, but the conflict is made more palpable by the antipathies between various members of his own family—between the hero and his family in general or his mother.[12] In particular, between his mother and the aunt who had stolen her husband,[13] between this aunt and the other aunts and uncles,[14] between the dead father and the mother, between the inhibited young cousin Mary and the mother and aunts. These characters are not treated satirically; even the stupidest uncle is allowed a certain human feeling and an inkling of truth outside himself. The old mother, who in a sense

[8] Son of Agamemnon and Clytemnestra, the avenger in Aeschylus' trilogy the *Oresteia*.

[9] Harry's recurrent nightmare is that he had pushed his wife overboard, as revealed in the first scene of the play.

[10] The Erinyes, spirits of punishment, avenged wrongs done to kindred, esp. murder within family or clan. Orestes is haunted by these Furies until he is reconciled to them as 'the kindly ones' ('Eumenides') in the last play of that title in the *Oresteia*. Eliot's Eumenides are revealed in *The Family Reunion* at the conclusions to I. ii and II. iii.

[11] Four Knights of King Henry II.

[12] Amy, Dowager Lady Monchensey.

[13] Agatha, a younger sister of Amy.

[14] Amy's other younger sisters Ivy and Violet; Col. the Hon. Gerald Piper and the Hon. Charles Piper, brothers of Amy's deceased husband.

has been a vampire to her son, yet compares favourably with the mother in Messrs Auden and Isherwood's *Ascent of F6*, who is almost a Freudian dummy.[15]

Technically the verse of this play is most successful, though some people have accused it of not being verse at all.[16] Mr Eliot has quite rightly avoided inserting any hunks of obvious prose; no prose-plus-verse play in recent times has as yet managed to be homogeneous. He has therefore had to contrive a versification elastic enough to be incantatory at one moment and to represent the banalities of conversation at another. This is a very considerable achievement. He uses his favourite devices—hypnotic repetition, antithesis, paradox, the overrunning of sentences from line to line, the simple and sharp but yet mysterious use of imagery:

> . . . the sobbing in the chimney
> The evil in the dark closet?[17]

And there are echoes from his previous poetry—'south in the winter',[18] '*You* don't see them, but I see them'[19] (the key line from the *Choephoroe* of Aeschylus).[20] It is foolish to cavil at these echoes when they are so well integrated into the present piece. Thus the scene between Harry and his Aunt Agatha is a reminiscence of *Burnt Norton*,[21] but is a magnificent presentation of the world of unfulfilled choices:

> I was not there, you were not there, only our phantasms
> And what did not happen is as true as what did happen,
> O my dear and you walked through the little door
> And I ran to meet you in the rose garden.[22]

[15] The Mother of Michael Ransom, hero of *The Ascent of F6* (1936), is stagily unveiled on a mountain summit at the play's climax (117)—an Oedipal resolution not well prepared for in the preceding action.

[16] e.g., Derek Verschoyle, review of the play at the Westminster Theatre, *Spectator* 162: 5778 (24 March 1939), 484, para. 1; anon. review of the book, 'Mr Eliot in Search of the Present: A Modern Verse Drama', *TLS*, no. 1938 (25 March 1939), 176, para. 2.

[17] *The Family Reunion*, I. ii, p. 32: *Collected Plays*, 68.

[18] I. ii, p. 12: *Collected Plays*, 57—echo of *The Waste Land*, i. 18.

[19] II. i, p. 24: *Collected Plays*, 64.

[20] *ὑμεῖς μὲν οὐχ ὁρᾶτε τάσδ', ἐγὼ δ' ὁρῶ*: *The Choephoroe* 1059. (T. G. Tucker in his edn. and trans. (1901) translated the next half-line, 'They harry me', which might have suggested to Eliot his protagonist's first name.)

[21] II. ii is reminiscent of *Burnt Norton* (1935), i.

[22] II. ii, p. 105: *Collected Plays*, 107.

Most of the characters speak at one time or another as if they were a chorus; this is one of the advantages of a poetic play. Further, Mr Eliot here has not introduced any external chorus (a disrupting influence on the modern stage) but on occasions (with a certain irony?) he makes the four stupidest characters step out of their proper parts and speak a commentary in unison.[23] I am not sure if this will succeed on the stage, but it is at least a hopeful experiment. It is probable, however, that this could have been dispensed with and that characters like Agatha could have been left to speak the commentary singly and still more or less in character.[24]

Lastly, this is a very moving play both as a whole and in its passing pictures, its ironic comments, its pregnant understatements, its bursts into liturgy. Witness Mary's criticism of Henry:

> . . . you attach yourself to loathing
> As others do to loving; an infatuation
> That's wrong, a good that's misdirected.[25]

Or Henry's comment on himself as a person that his family has conspired to invent.[26] Or one of his first remarks on remeeting them after eight years: 'You all look so withered and young.'[27] Or his mother's dying words: 'The clock has stopped in the dark.'[28] Or the brilliant reminiscences of a neurotic childhood.[29] Or Henry's indication of his apparently eccentric conduct:

> In a world of fugitives
> The person taking the opposite direction
> Will appear to run away.[30]

Mr Eliot's own poetry may appear to be taking the opposite direction, but the reader of this play cannot, I think, object to it, as he could to *The Waste Land*, that it is essentially defeatist; it embodies a sincere belief and a genuine courage.

[23] Ivy and Violet, Gerald and Charles, Harry's aunts and uncles, form a family chorus in four scenes of the play: I. i, I. iii, II. i, II. iii.

[24] Eliot dispensed with the device of chorus in his last plays.

[25] I. ii, p. 57: *Collected Plays*, 81. From this point on, MacNeice changes the name of Harry to the formal Henry.

[26] I. i, p. 27: *Collected Plays*, 65.

[27] I. i, p. 25: ibid., 64.

[28] II. iii, p. 126: ibid., 119.

[29] Chiefly in Harry's reminiscences to Mary in I. ii.

[30] II. iii, p. 110: *Collected Plays*, 110—spoken by Agatha, not Harry.

The Poet in England To-day: A Reassessment[1]

For some years before this war English literary circles had been largely obsessed with sociology. The younger poets in reaction against a decadent individualism regarded the poet as primarily a community spokesman; as there was no healthy community in existence he had to be the mouthpiece of a future regime. But the protagonists of the regime were already with us—the Workers. This word 'Worker'—and the other word 'Proletariat'— became heavily overcharged with mysticism. It was not recognized that in a corrupt society the Workers are infected with the generally prevailing disease; proletarians, it was assumed, were exempt from original sin.[2] This led to a sentimental self-abasement on the part of many intellectuals. Edward Upward, in an essay in *The Mind in Chains*, asserted that no one nowadays can possible write anything of any value unless he is taking an active part in 'The Workers' Movement'.[3] Upward, who was the early inspiration of Auden and Isherwood, also wrote a novel depicting the introspective fantasies of a neurotic young man who after two hundred pages sees the light and goes off 'to find a worker'—as if that would solve everything.[4] The Proletariat had become the *Deus ex Machina*.

This, in my opinion, was no compliment to the Proletariat; we all know the worth of gods-in-machines. The intellectual assumed—rightly, I think—that it was no longer his vocation to sit face inwards in a corner weaving from his own intestines a spider's web of private fantasy. So he turned face outwards and

[1] *New Republic* 102: 13 (25 March 1940), 412–13; subtitle added from the holograph manuscript (Humanities Research Center Library, University of Texas, Austin). MacNeice published his *Selected Poems* (Faber) in March 1940.

[2] Cf. title of MacNeice's review of Eliot's *The Family Reunion* (May 1939), also in the *New Republic*, printed above.

[3] Edward Upward, 'A Marxist Interpretation of Literature', *The Mind in Chains*, ed. C. Day-Lewis (1937), 53—read by MacNeice in July 1937—*I Crossed the Minch*, 105.

[4] End of Upward's *Journey to the Border* (1938): the protagonist, an anon. tutor, meets his double (voice of conscience) who advises him to join the workers' movement (213–14); he finds a factory worker at a race course to talk to (225–33), and then makes up his mind to join the workers (255–6)—extracts in *New Writing*, nos. 1 (Spring 1936) and 3 (Spring 1937).

seeing the Proletariat moping in another corner stampeded across to join it. The Proletariat was the only vehicle for future progress; drawing from this the fallacious inference that it was the only vehicle for present truth, the intellectual, with astonishing *naïveté*, and at the same time proud of his self-sacrifice, sloughed off his critical faculty and sat at the feet of the Proletariat saying, 'Feed me infallible truths.' As for poetry, it was decreed that from now on this must come only from 'the people' and be written for 'the people', who were equated with this contemporary, diseased, heterogeneous proletariat. The mathematicians also put away their higher mathematics, saying 'Feed us a loaf and a loaf makes two.' Or, if they didn't, they should have.

Thus a very highbrow young novelist got up at an anti-fascist meeting and said that no writer nowadays ought to have either a will or a conscience or a critical faculty of his own: 'the Workers' must do all his willing and moralizing and criticizing for him. In the same way certain sophisticated poets carefully dragooned themselves into the narrow (but easy) paths of slogan-poetry—'up the barricades!' Auden and Isherwood wrote a bad play, *On the Frontier*,[5] in the belief that writing down in literature is the best way up in life. Stephen Spender toured England lecturing on 'The Artist and Society' and trying to make Picasso a Marxian evangel.[6] It was possible to suspect that some of the intellectuals took up the cult of the Proletariat in the same way that W. B. Yeats took up the cult of his spirits (who came, in their own words, 'to give him metaphors for his poetry').[7] For, whatever is the case with other men, a poet's approach—even if he is Mayakovsky[8]—is always essentially personal.

Man's deference to any logic of black-and-white, of all-or-nothing, is probably due to his basic illogicality; he just cannot cope with the world in colour. He refuses to distinguish conditions from causes. He cannot recognize the importance of the Economic Factor without trying to split the atom with it. A leopard cannot

[5] See MacNeice's review (Nov. 1938) above, page 101.

[6] The substance of this lecture was published in a Hogarth pamphlet *The New Realism: A Discussion* (1939) by Spender; see also Sir Stephen's article, 'Picasso's Guernica', *New Statesman and Nation*, NS, 56: 399 (15 Oct. 1938), 567–8. Picasso joined the Communist Party to support the republican party in the Spanish Civil War.

[7] Introduction to Yeats's *A Vision* (1937, 1956), 8: 'We have come to give you metaphors for poetry'.

[8] See n. 53 to the article 'Subject in Modern Poetry' (Dec. 1936) above.

change his spots but a man, having divided the world into sheep and goats, can always become a sheep overnight. Socially or politically—using these words in a narrow sense—this may be a good thing; it has seemed up till now that men in the mass must have their issues simple, i.e. false. But society is something more than men in the mass, just as a man is something more than a conglomeration of cells. The fact that soul is inseparable from body does not disprove soul. The fact that man's various expressions of freedom can all be referred back by the Marxian historian to an inexorable causality—and by the psychoanalyst to another, but equally inexorable, causality—does not disprove freedom.

Art, though as much conditioned by material factors as anything else, is a manifestation of human freedom. The artist's freedom connotes honesty because a lie, however useful in politics, hampers artistic vision. Systematic propaganda is therefore foreign to the artist in so far as it involves the condoning of lies. Thus, in the Spanish Civil War some English poets were torn between writing good propaganda (dishonest poetry) and honest poetry (poor propaganda).[9] I believe firmly that in Spain *the balance of right* was on the side of the government; propaganda, however, demands either angels or devils. This means that in the long run a poet must choose between being politically ineffectual and poetically false. For the younger English poets the choice has now been simplified. A poet adopts a political creed merely as a means to an end. Recent events having suggested that there are too many slips between certain means and certain ends, the poet is tending to fall back on his own conscience.

This does not mean a retreat to the Ivory Tower or to purely private poetry. To assume that it does is bad logic. If the artist declines to live in a merely political pigeonhole, it does not follow that he has to live in a vacuum. Man is a political animal, not a political cog. And to shun dogma does not mean to renounce belief. I. A. Richards wrote in the twenties, in *Science and Poetry*: 'It is never what a poem says which matters, but what it *is*.'[10]

[9] For the latter, see *Poems for Spain*, ed. Stephen Spender and John Lehmann, with intro. by Spender (1939), and with sect. xxiii from MacNeice's *Autumn Journal* 1938 (1939) included under the title 'Remembering Spain', 97–100.

[10] (1926, rev. 1935), 31 (reissue: *Poetries and Sciences* (1970), 33). Cf. Archibald MacLeish's widely-quoted poem 'Ars Poetica' (1926) ending 'A poem should not mean / But be': *New & Collected Poems, 1917–1976* (1976), 106–7.

This may be a half-truth but it is nowadays worth emphasizing, the poets of the thirties having been so much concerned with what they were saying, and to whom, and with whose approval. Not that I accept Richards's complete severance of poetry from beliefs; a poem flows from human life with which beliefs are inevitably entangled. But to let beliefs monopolize poetry is to be false not only to poetry but to life, which is itself not controlled by beliefs. Some of the poets who renounced the Ivory Tower were ready to enter a Brazen Tower of political dogma; where the Ivory Tower represents isolation from men in general, the Brazen Tower represents isolation from men as individuals (witness the typical entowered politician) and also from oneself as an individual. Bad logic demanded a choice between the Towers, but salutary self-deceit allowed many of the Brazen school to leave the door open. The impact of the war with its terrible threat of genuine spiritual imprisonment has brought them again out of doors. The poet is once more to be a mouth instead of a megaphone,[11] and poetry, one hopes, is to develop organically from the organic premisses of life—of life as it is lived, not of life when it is dried into algebra.

Not Tabloided in Slogans[1]

Another Time. Poems by W. H. Auden. (Random House. $2.00.)

Poets in England lately have been changing their position, recent events having suggested that their position was unsound. The poet has become at the same time more humble and more arrogant, being less ready now to take up the role of crusader but ready once more to put his own conscience above any external dogma. In the case of W. H. Auden, the most gifted and the most exciting poet of his generation, the change seems to have coincided with his settling last year in America. At one time the advocate of

[11] Cf. Auden's poem 'In Memory of W. B. Yeats' (Feb. 1939), *The English Auden*, 242: 'poetry . . . survives, / A way of happening, a mouth.'

[1] *Common Sense* 9: 4 (April 1940), 24–5.

directly political writing, he now regards politics as only too likely to corrupt a poet's integrity. This does not mean that he has retreated to the Ivory Tower. To cease to be politically propagandist does not mean that you cease to be socially conscious. Auden knows as well as anyone that the individual in a vacuum is a deficient individual.

His very remarkable new book of poems appears to embody a semi-mystical philosophy of life, girded with conceptions from the psychologists. 'In Memory of Sigmund Freud' is significantly one of the best poems and suggests that he is now banking on salvation from within rather than salvation from without. On the whole the new philosophy is difficult to define, especially in its treatment of the problem of Evil, for Auden is primarily a poet and only a philosopher incidentally; ideas strike him—as they did Donne—emotionally and so are hardly detachable from their context, from the mood of any particular poem. This knack of fusion means that when he writes allegorically, which he does more and more often, he avoids—as Bunyan did in *Pilgrim's Progress*—the sterility of common allegory which, remaining algebra, falls short of incarnation.

This book displays once more his astonishing versatility, ranging from the most elegant kind of lyric to 'occasional' poetry which has the easy tone of very intelligent table-talk. I am not sure that I like his rather flashy satirical ballads (e.g. the sadistic 'Miss Gee') though I fully concede the poet's right to be satirical or vulgar or frivolous when he feels like it. A very quiet piece on the other hand, 'Musée des Beaux Arts', seems to me true, exact and moving. And, while some of the poems about people and places—Matthew Arnold and Edward Lear, for example, or Oxford and Dover—are a little too pat, too much written to formula, there is plenty of beauty and significant writing in this book to outweigh them.

Some of Auden's fans are complaining that he has sold the pass. If he were merely—as he may think he is—plumping on personal relationships (in the narrow sense), that might be so. What he is really doing is recognizing—what the politician forgets—that the world is a world of persons. He is reaffirming the Rights of Man, those particular rights which cannot be tabloided in slogans. His psychological dialectic is not defeatist because it recognizes that

Necessary acts are done
By the ill and the unjust[2]

That man's course is a winding one does not disprove it is a course.

Yeats's Epitaph[1]

Last Poems and Plays, by W. B. Yeats. (New York: The Macmillan Company. 131 pages. $1.75.)

During the last ten years, Yeats has had more bouquets from the critics than any other poet of our time. It was refreshing to see these critics and also many of the younger poets committing themselves to enthusiasm for an older contemporary; their praise, however, was sometimes uncritical and sometimes, on a long-term view, injurious to its subject. There were reviewers who felt Yeats was a safe bet—safe because he was an exotic; anyone can praise a bird of paradise but you have to have some knowledge before you go buying Rhode Island Reds.[2] There is a double point that needs making—first that Yeats was not so exotic as is popularly assumed, second that on the whole his exoticism was not an asset but a liability. He was partly aware of this himself; in his middle period he fought clear of the dead hand of Walter Pater[3] and deliberately set out to make his poetry less 'poetic' and in his later years (the years when he was a devotee of Balzac) he paid at least lip-homage to the principle of 'Homo sum. . . .'[4] His failure fully to practise this principle was due to a constitutional inhumanity.

I say this in honour to his memory. If you believe a man was a genius, it is an insult to him to ignore his deficiencies and peculiarities. One of the most peculiar poets in our history, Yeats

[2] 'The Riddle', st. 2: *Collected Poems*, ed. Edward Mendelson (1976), 204.

[1] *New Republic* 102: 26 (24 June 1940), 862–3. Yeats had died 28 Jan. 1939.

[2] A breed of New England poultry producing brown-shelled eggs.

[3] Walter H. Pater (1839–94), stylist of elaborate English prose influential among 1890s æsthetes of whom Yeats was one.

[4] *Homo sum, humani nihil a me alienum puto* ('I am a man, and consider nothing human alien to me'): Terence, *Heauton Timoroumenos* (*The Self-Tormentor*, 163 BC) 77.

was also extraordinarily lacking in certain qualities which the greater poets usually possess; in so far as he achieved greatness it proves, not the power of inspiration or any other such woolly miracle—all that it proves is the miracle of artistic integrity. For this was a quality he possessed even though as a man he may sometimes have been a fraud. His more naïve enemies regard him as knave or fool all through—at best as a 'silly old thing'; his more naïve admirers regard him as God-intoxicated and therefore impeccable. It is high time for us to abandon this sloppy method of assessment; if poetry is important it deserves more from us than irresponsible gibes on the one hand or zany gush on the other.

Take Yeats's two passions—Ireland and Art. We have to remember that, in regard to both, his attitude was conditioned by a comparatively narrow set of circumstances and that, in judging his services to Ireland and Art, we shall be very shortsighted if we reapply his own heavily blinkered concepts of either; it is a lucky thing for the artist that his work usually outruns his ideology. Yeats talked a good deal about magic and beauty and mysticism, but his readers have no right to gabble these words like parrots and call what they are doing appreciation. Beauty is *not* the mainspring of poetry and, although a few poets have been genuine mystics, Yeats, unlike his friend AE,[5] was certainly not one of them; he had what might be called a mystical sense of value, but that is a different thing and a thing which perhaps for *all* artists is a *sine qua non*.

Yeats's poetry reached its peak in *The Winding Stair* (1933). The *Last Poems* now published represent the Indian summer of his virile, gossipy, contumacious, arrogant, magnificently eccentric old age. Although the book as a whole certainly lacks the depth and range of *The Tower* or *The Winding Stair*, although the septuagenarian virility is sometimes too exhibitionist, although he overdoes certain old tricks and falls into needless obscurities, and although the two plays here included are flat failures,[6] there is still enough vitality and elegance to compensate for certain disappointments. Few poets in English literature have been able

[5] Pseud. (derived from 'aeon') of George W. Russell (1867–1935), Irish mystical poet and friend of Yeats.

[6] *Purgatory* and *The Death of Cuchulain* 1938–9 (1939) are no longer considered failures in drama, although they may require a special audience.

to write lyrics after thirty-five; the astonishing thing about Yeats is that he remained essentially a lyric poet till the last. Even the enormous cranky pseudo-philosophy of *A Vision* only served as an occasion for further lyrics.[7] Yeats's ingredients became odder and odder but, because they were at least dry and hard, they helped him to assert a joy of life which was comparatively lacking in his early Celtic or Pre-Raphaelite twilights. The great discovery of the later Yeats was that joy need not imply softness and that boredom is something more than one gets in dreams. Axel has been refuted;[8] 'Hamlet and Lear are gay.'[9]

Ireland is very prominent in these last poems. Yeats had for a long time regarded the essential Ireland as incarnate in the country gentry and the peasantry, his ideal society being static and indeed based upon caste. The Irish 'Troubles', however, evoked in him an admiration, even an envy, for the dynamic revolutionary.[10] His thought, in assimilating this element, became to some extent dialectical; he began to conceive of life as a developing whole, a whole which depends upon the conflict of the parts. He even began to write in praise of war, a false inference from a premiss which is essentially valid—

> . . . when all words are said
> And a man is fighting mad,
> Something drops from eyes long blind,
> He completes his partial mind. . . . [11]

Physical violence being a simple thing, the Yeats who honoured it took to writing in ballad forms, while the contemplative Yeats continued to use a grand rhetorical manner and a complex inlay of esoteric ideas and images; there are good examples of both in this book. There are also, as in the preceding volumes, a number of poems about himself and his friends; once again he goes around

[7] Many of Yeats's poems written after his marriage in 1917 were inspired by his constructed 'pseudo-philosophy' of *A Vision* (1925; 1937, 1956), esp. from 'The Double Vision of Michael Robartes' (1919) to 'Byzantium' (1930) to 'The Gyres' (1936–7).

[8] Refutation of the life-denying French symbolist drama *Axël* (1890), a play by Villiers de l'Isle Adam (1838–89) which ends with a Wagnerian love-death.

[9] 'Lapis Lazuli', l. 16.

[10] The 'Troubles' of 1916–22 between Britain and Sinn Fein (Irish republican party) inspired Yeats to write such masterful poems as 'Easter 1916', 'Meditations in Time of Civil War' 1921–22, and 'Nineteen Hundred and Nineteen' 1919–22.

[11] 'Under Ben Bulben' (Feb. 1939), iii. 3–6.

with a highly coloured spotlight; it is amusing to turn from one of these poems, 'Beautiful Lofty Things', which contains a reference to a public banquet in Dublin, to George Moore's account of the same banquet in *Ave*.[12] The most revealing poem in this book is 'The Circus Animals' Desertion', where Yeats with admirable ruthlessness looks back on his various elaborate efforts to project himself on to the world—Celtic legend, Maud Gonne, symbolic drama. In this excellent and moving poem a self-centred old man rises above his personality by pinning it down for what it is.[13]

The Tower that Once[1]

Mrs Woolf, in her article 'The Leaning Tower',[2] looks forward to a classless society which will give to writers 'a mind no longer crippled, evasive, divided. They may inherit that unconsciousness which . . . is necessary if writers are to get beneath the surface, and to write something that people remember when they are alone.' With this general aim or hope I sympathize. 'Literature', I would agree with her, 'is no one's private ground.' I find it, therefore, both inconsistent and unjust that she should dismiss not only so lightly, but so acidly—as 'the embittered and futile tribe of scapegoat hunters'—that group of younger writers[3] who during the Thirties made it their business to stigmatize those all too present evils which Mrs Woolf herself considers evil and to open those doors which she herself wants opened. She seems to understand these junior colleagues of hers no better than Yeats

[12] For an account of Standish O'Grady (1846–1928), 'father of the Irish literary revival', addressing a public dinner for the Irish Literary Theatre at the Shelbourne Hotel (given by the *Daily Express*, Dublin), see Moore's *Ave* (1911, 1919), chap. 4; cf. Yeats's *Dramatis Personae* (1935), sect. xi, *Autobiographies* (1955), 423–4.

[13] Conclusion to the poem: ' . . . now that my ladder's gone, / I must lie down where all the ladders start, / In the foul rag-and-bone shop of the heart.' Cf. MacNeice's article 'In Defence of Vulgarity', *Listener* 18: 468 (29 Dec. 1937), 1407–8—to be published in the second volume of selected prose.

[1] *Folios of New Writing* 3 (Spring 1941), 37–41.

[2] *Folios of New Writing* 2 (Autumn 1940), 11–33; reprinted in *The Moment and Other Essays* (1947); collected in *Collected Essays* by Virginia Woolf, 4 vols. (1966, 1967), ii. 162–81.

[3] 'Day Lewis, Auden, Spender, Isherwood, Louis MacNeice and so on': Woolf, 'The Leaning Tower', *Collected Essays*, ii. 170.

understood Eliot.[4] This mutual misunderstanding of the literary generations is one of the evils of our times; my own generation has too often been unjust to its immediate predecessors.

Mrs Woolf's literary history is over-simplified. She writes of the social divisions of the nineteenth century: 'the nineteenth-century writer did not seek to change those divisions; he accepted them. He accepted them so completely that he became unconscious of them.' Confining ourselves to our own literature and leaving aside all foreigners—the Russians, for example, or Zola—we might ask her what about Shelley whom she herself has mentioned by name. Or what about Wordsworth whose early inspiration was 'Nature' admittedly but Nature harnessed to a revolutionary social doctrine? And even with the great Victorians 'All's well with the world' was not their most typical slogan; is *In Memoriam* a poem of placid or unconscious acceptance? 'Life was not going to change', writes Mrs Woolf; Tennyson said something different in 'Locksley Hall'.[5] And what about William Morris? Or Henry James, for whom (according to Mrs Woolf) as for (according to her too) his predecessors the social barometer was Set Fair for ever? Mr Spender put forward a different, but at least as plausible a view of James in *The Destructive Element*.[6]

Mrs Woolf assumes that a period of great social and political unrest is adverse to literature. I do not think she produces adequate evidence for this; we could counter with the Peloponnesian War, the factions of Florence in the time of Dante, the reign of Queen Elizabeth, the Franco-Prussian War; but, even if she is right, she should not attack my generation for being conditioned by its conditions. Do not let us be misled by her metaphor of the Tower. The point of this metaphor was that a certain group of young writers found themselves on a leaning tower; this presupposes that the rest of the world remained on the level. But it just didn't. The whole world in our time went more and more on the slant so that no mere abstract geometry or lyrical uplift could cure it.

[4] Yeats attacked the prosiness of Eliot's verse in his BBC broadcast, 11 Oct. 1936, 'Modern Poetry', in *Essays 1931 to 1936* (1937); collected in *Essays and Introductions*, 499.

[5] 'Not in vain the distance beacons. Forward, forward let us range, / Let the great world spin forever down the ringing grooves of change': Tennyson, 'Locksley Hall' (1842), ll. 181–2.

[6] See MacNeice's review of Spender's book, 'Modern Writers and Beliefs' (May 1935), above.

When Mrs Woolf accuses the Thirties writers of 'flogging a dead or dying horse because a living horse, if flogged, would kick them off its back,' her point seems to me facile. No doubt we spent too much time in satirizing the Blimps, but some of those old dead horses—as this war shows every week—have a kick in them still. And the ruling class of the Thirties, the people above the Blimps, our especial *bête noir* or *cheval noir*, did manage to kick us into the jaws of destruction. But it remains to be seen who will be proved to have died; we'll hope it was the horse.

She proceeds to a surprising sentence: 'How can a writer who has no first-hand experience of a towerless, of a classless society create that society?' How can a larva with no first-hand experience of flight ever grow wings? On the premisses implied in this sentence human society is incapable of willed or directed change. Because, quite apart from the intelligentsia and the privileged classes, there is nobody in the whole population of Great Britain who has had firsthand experience of that kind of society which nearly everybody needs. Mrs Woolf is making the same mistake as some of the very writers she is attacking. For some of those writers were hamstrung by modesty. It all, they said to themselves, depends on the proletariat. And in a sense they were right. But they were wrong to assume that the proletariat itself knew where it was going or could get there by its own volition. These intellectuals tended to betray the proletariat by professing to take all their cues from it.

Mrs Woolf deplores the 'didacticism' of the Thirties. But (1) if the world was such a mess as she admits, it was inevitable and right that writers should be didactic (compare the position of Euripides),[7] (2) she assumes that this writing—especially the poetry—of the Thirties was solely and crudely didactic—which it was not. She makes an inept comparison between a morsel of Stephen Spender and a morsel of Wordsworth as exemplifying 'the difference between politician's poetry and poet's poetry': this ignores the fact that the great bulk of Wordsworth is pamphleteering and that Spender's poetry is pre-eminently the kind—to use her own words—'that people remember when they are alone.' Politician's poetry? Look at Spender's professedly

[7] The Peloponnesian War was reflected in Euripides' so-called 'political' plays, such as *The Suppliant Women* (*c.*420 BC) and the *Andromache* (*c.*419 BC).

political play, *Trial of a Judge*: it failed as a play just because it was not 'public' but rather a personal apologia; it displeased the Communists just because it sacrificed propaganda values to honesty.[8]

It is often assumed by the undiscriminating—among whom for this occasion I must rank Mrs Woolf—that all these writers of the Thirties were the slaves of Marx, or rather of Party Line Marxism. Marx was certainly a most powerful influence. But why? It was not because of his unworkable economics, it was not because of the pedantic jigsaw of his history, it was because he said: '*Our job is to change it.*' What called a poet like Spender to Marx was the same thing that called Shelley to Godwin and Rousseau. But some at least of these poets—in particular Auden and Spender—always recognized the truth of Thomas Mann's dictum: 'Karl Marx must read Friedrich Hölderlin.'[9] Even an orthodox Communist Party critic, Christopher Caudwell, in his book *Illusion and Reality*, insisted (rightly) that poetry can never be reduced to political advertising, that its method is myth and that it must represent not any set of ideas which can be formulated by politicians or by scientists or by mere Reason and/or mere Will—it must represent something much deeper and wider which he calls the 'Communal Ego'.[10] It is this Communal Ego with which Auden and Spender concerned themselves.

Politician's poetry? Yes, there was some of it; and some of it was bad. Rex Warner, for example, lost his touch when he turned from birds to polemics. Day-Lewis's social satire cannot compare with his love lyrics. Auden and Isherwood's *On the Frontier* was worse than a flop. Mr Edward Upward ruined his novel, *Journey to the Border*, with his use of the *Deus ex Machina* —i.e. 'the Workers'—at the end.[11] But these mistakes are nothing to their achievements and it is grotesque to dismiss someone like Auden

[8] Cf. MacNeice's account of Spender's brief time in the Communist Party centring around *Trial of a Judge* (1938) in *The Strings are False*, 166–8.

[9] 'Karl Marx den Friedrich Hölderlin gelesen hat', from essay 'Goethe und Tolstoi' (1922) in *Gesammelte Werke*, 13 vols. (1960–74), ix. 170. Note MacNeice's emphasis or mistranslation.

[10] 'Caudwell' (pseud. of Christopher St John Sprigg, 1907–37), *Illusion and Reality: A Study of the Sources of Poetry* (1937, 1946), 151 *et passim*.

[11] Upward, *Journey to the Border* (1938) unrealistically 'liberates' his tutor-protagonist by having him join 'the workers' movement' at the end: cf. above MacNeice's article 'The Poet in England To-day' (March 1940), para. 1, with n. 4.

as a mere 'politician's poet' and an ineffectual one at that; was it not Auden who repudiated the Public Face in the Private Place? It is carrying the Nelson eye[12] too far to pretend that Auden and Spender did not bring new life into English poetry and what was more—in spite of what Mrs Woolf says about self-pity—a new spirit of hopefulness (see some of Spender's early lyrics). As for the novel, Mrs Woolf suggests that, whereas her own generation could create objective character and colour, her successors can manage nothing but either autobiography or black and white cartoons. I would ask the reader—with no disrespect to *Mrs Dalloway*, a book that I like very much—to compare Mrs Woolf's *Mrs Dalloway* with Mr Isherwood's *Mr Norris*.[13]

Self-pity? Of course our work embodied some self-pity. But look at Mrs Woolf's beloved nineteenth century. 'Anger, pity, scapegoat beating, excuse finding'—she intones against the poor lost Thirties; you find all those things—in full measure and running over—in the Romantic Revival and right down from *Manfred* or Keats's Odes through Tennyson and Swinburne and Rossetti to the death-wish of the Fin de Siècle and even to Mr Prufrock. My generation at least put some salt in it. And we never, even at our most martyred, produced such a holocaust of self-pity as Shelley in *Adonais*.[14]

Mrs Woolf deplores our 'curious bastard language', but I notice that in the next stage of society and poetry she looks forward to a 'pooling' of vocabularies and dialects. Just one more inconsistency. And Shakespeare wrote in a bastard language too.

This is no occasion to put forward a *Credo* of my own, but I would like to assure Mrs Woolf (speaking for myself, but it is true of most of my colleagues) that I am not solely concerned with 'destruction'. Some destruction, yes; but not of all the people or all the values all the time. And I have no intention of recanting my past. Recantation is becoming too fashionable; I am sorry to see so much self-flagellation, so many *Peccavis*, going on on the

[12] i.e., a blind eye.

[13] Isherwood's lively stories, *Mr Norris Changes Trains* (1935), *Sally Bowles* (1937), *Goodbye to Berlin* (1939), are collections of episodes rather than tightly integrated novels like Woolf's *Mrs Dalloway* (1925).

[14] Shelley's *Adonais* (1821) on the death of Keats is actually a sentimentalized elegy commiserating the poet writing it.

literary Left. We may not have done all we could in the Thirties, but we did do something. We were right to throw mud at Mrs Woolf's old horses and we were right to advocate social reconstruction and we were even right—in our more lyrical work—to give personal expression to our feelings of anxiety, horror and despair (for even despair can be fertile). As for the Leaning Tower, if Galileo had not had one at Pisa, he would not have discovered the truth about falling weights. We learned something of the sort from our tower too.

The Elusive Classics[1]

Virgil: *The Eclogues and the Georgics*. Translated into English verse by R. C. Trevelyan. (Cambridge University Press. 7*s*. 6*d*.)

As words cannot do everything, poets have often to compromise. Translators, *a fortiori*, must nearly always compromise—for each language has its own deficiencies as well as its own capacities. Thus, where the Latin language almost calls for certain things to be said in it, the English language will sometimes hardly allow them to be said; this especially applies to Horace and Virgil. What is the translator to do when confronted with such poets whose literal—or prose—meaning counts for little in proportion to their total—or poetic—meaning? This latter meaning is brought out in the original by all kinds of formal subtleties to which in English verse we cannot do more than approximate. If the translator plumps for 'meaning' in the narrow sense, i.e. for literal accuracy, he is liable to turn out something as dull as a school crib—and, in the long run, as misleading. If, on the other hand, he plumps for the feel of the poetry as poetry, and frees himself from literal accuracy in order to 'recreate' the 'spirit' of his original, he most often either creates something quite alien to that original or, because he is himself uncreative, fails to create anything at all. In English we have had during the last century many shocking examples of this kind of free translation.

[1] *New Statesman and Nation* 29: 724 (5 May 1945), 293.

Nowadays there is among translators a reaction from freedom in favour of 'faithfulness'—but of a faithfulness that will go beyond the school crib. Beginning with the literal content, they then, while adding to, or subtracting from, that content as little as possible, try to salvage *via* their *presentation* of it—i.e., by phrasing, word-order, rhythm, vowel-pattern, etc.—those elements which in the original heightened the content into poetry. It is to this school that Mr Trevelyan belongs, and in these latest translations, having a genuine love for his author and being a man of integrity, he does give something more than the merely narrow meaning, and that something more is, so far as it goes, Virgilian. He is not very daring with words, and the virtues of his method are largely negative; he has obviously been at pains to avoid false diction, stunting, and obviously wrong forms. Negatively, then, he is on the right track. This is not such faint praise as it may sound; surrounded as we are by libraries of viciously translated classics, our first duty as translators is to avoid the vices of our predecessors. If, in avoiding those, we produce something that is dignified, though maybe bare, and rings—if but faintly—true, we have done something worth while. Thus, Mr Trevelyan's *Georgics*, though not great translations, are worth while.

I say 'not great', because, to my feeling, he lacks two things—creative power and the poetic sense of touch. Thus, while avoiding the grosser errors of taste, he technically commits certain errors of judgment—for instance, in his metrics. In translating the *Eclogues* he takes as his norm 'an unrimed verse of seven, and occasionally eight, accents', which he compares misleadingly to 'the normal half-stanza of the English ballad',[2] but which more resembles the loose versification of Blake's prophetic books. Mr Trevelyan chose this form because, being about the same length as an hexameter, it allowed him 'to translate line for line with very little omission or expansion'—a good reason; also because this metre 'has at least the merit of swiftness of movement'—another good reason, if it were true.[3] Unfortunately, the movement of his *Eclogues*, like the movement of Blake's prophetic books, is, on the whole, shambling rather than rapid. In the *Georgics*, on the other hand, he gives up the attempt at a line-for-line translation, and falls

[2] Trevelyan, 'Introduction' to *The Eclogues* (1944), 3.

[3] Loc. cit.

back on blank verse as being 'the subtlest and most plastic' of English metres, explaining that in Virgil 'mastery of movement comes first in importance'.[4] But his version, for all its other merits, does not show mastery of movement. Inevitably—and rightly—he varies his rhythms, inverting the stresses or inserting extra syllables, but these variations—to my ear—do not come naturally.

A verse translation, ideally, should be what Dr Johnson demanded of Wit: 'at once natural and new'.[5] Mr Trevelyan lacks that novelty which will make an old theme live for a modern reader. It is instructive here to compare him with Cecil Day-Lewis, whose translation of the *Georgics* was published in 1940. Many readers, who feel at home with blank verse; and who dislike colloquial diction (let alone slang), will prefer Mr Trevelyan's version, but for me Mr Day-Lewis, being himself more creative, is more recreative of Virgil in rhythm, texture and phrasing. To begin with, he was happier in his choice of 'a rhythm based on the hexameter, containing six beats in each line, but allowing much variation of pace and interspersed with occasional short lines';[6] this achieves what Mr Trevelyan attempted but failed to achieve in his *Eclogues*.

Quotations will illustrate the difference between these two versions. For Virgil's 'Arctos Oceani mentuentis aequore tingui', where Trevelyan has 'the Bears who dread to plunge 'neath Ocean's plain,' Day-Lewis, while losing somewhat the meaning of *aequor*, retains the advantage of a line for a line, and gets both a more appropriate rhythm *and*, I think, a more vivid picture in 'The Bears that are afraid to get wet in the water of Ocean.'[7]

Compare again, for 'humescunt spumis flatuque sequentum', Trevelyan's 'with the foam And breath of their pursuers they are wet,' with Day-Lewis's 'The drivers are wet with the spindrift breath of the horses behind them.'[8] Lastly, as an example of what can be done with one word, where Virgil says of a swarm of bees

[4] Ibid., 'Introduction' to *The Georgics* (1944), 36.
[5] Samuel Johnson, 'Cowley', para. 55, in *Lives of the English Poets* (1779–81).
[6] Day-Lewis, 'Foreword', *The Georgics of Virgil* (1940), 8.
[7] Virgil, *Georgics* i. 246; Trevelyan, 47; Day-Lewis, 23.
[8] *Georgics* iii. 111; Trevelyan, 81; Day-Lewis, 58.

'glomerantur', Trevelyan has 'conglobed' (which sounds stilted), but Day-Lewis has 'scrimmage'—which sounds like bees.[9]

The above comparisons will, I hope, not appear unfair to Mr Trevelyan; he is here, after all, up against one of the most technically accomplished poets now writing in English. I might add that the Virgilian glamour often enough eludes both of them, e.g., in the line 'tantus amor florum et generandi gloria mellis.'[10] This sort of thing will always be there to be chased. All honour to those who, like Mr Trevelyan, chase it.

Introduction to *The Golden Ass of Apuleius*[1]

In the classical forms of our schools Apuleius, if mentioned at all, is usually dismissed as a decadent, in the form-master's eyes his matter being frivolous and indecent and his Latin not being Latin (the literature, of course, is assumed to end with Tacitus who died round AD 120 about the time Apuleius was born). Our teachers of classics have always been fond of the chopper; its use here deprives their pupils not only of much-needed light on the Antonine Age but also of a rare æsthetic experience. For Apuleius, if a freak, was also a great artist. At a time when original writing had lapsed into Greek (witness Lucian and Marcus Aurelius) he made a new drive on Latin. Literature had for some time been decentralized, the Silver Age writers having mostly come from Spain. Now it was Africa's turn; Apuleius was to be followed there by Tertullian and the early Church Fathers (who detested him). His education was cosmopolitan—Carthage, Athens, Rome—and his genius correspondingly eclectic. It is hard to find a writer who combines such dissimilar qualities—elegance

[9] *Georgics* iv. 79; Trevelyan, 101; Day-Lewis, 79.

[10] *Georgics* iv. 205, rendered as 'Such is their love for flowers, their pride in producing honey' by Day-Lewis, 83, and as 'so great their love of flowers, / So proud their glory in begetting honey' by Trevelyan, 105.

[1] *The Golden Ass of Apuleius*, trans. William Adlington 1566, with an Introduction by Louis MacNeice, in the Chiltern Library (London: John Lehmann, 1946), v–ix. In Nov. 1944 MacNeice had produced two Apuleian adaptations for broadcast: *The Golden Ass* and *Cupid and Psyche*.

and earthiness, euphuism and realism, sophistication and love of folk-lore, Rabelaisian humour and lyrical daintiness, Platonism and belief in witchcraft, mysticism and salty irony. In spite of his wild burlesques of the subject he obviously believed in magic, and for all his worldliness his Isis-worship was genuine. Still, he was predominantly an artist; the *Golden Ass* is not just a mixture but a blend.

This astonishing novel, originally entitled the *Metamorphoses* and belonging to the class of episodic and picaresque writings known as 'Milesian tales', is the work of a conscious innovator. Thoroughly trained as a rhetorician Apuleius avoids that fatal frigidity which rhetoric induced in so many Latin writers; he keeps his eye on the object and fulfils the promise in his first sentence to 'delight your kindly ears with a pleasant history'—*auresque tuas benivolas lepido susurro permulceam. Susurrus* is a word used for the humming of bees while *lepidus* (the nearest translation is probably 'charming') is repeatedly applied by Apuleius to stories of unabashed brutality like that in Book II of the man whose nose was eaten off by a witch. This gives us a clue to his mentality; a completely self-conscious intellectual, he yet retained the peasant's delight in everything violent and horrific. But his horror stories, like his obscenities, are clear-cut and hard—one might almost say clean—and free from nordic morbidity. It is not surprising that some of his yarns were borrowed by Cervantes and Boccaccio; nor that Adlington's English translation, published in 1566, should have been so popular with the tough Elizabethans who were not so far as we are from the Mediterranean tradition and were still immune from puritanism, subjectivity, and even conventional propriety.

Especially refreshing to one brought up on 'classical' Latin is Apuleius' virile yet delicate sense of humour. Here only Petronius rivals him, though the quintessence of such humour appears in certain poems of Catullus. All three writers, throwing off the toga of literary conventions, seem to have gone back into the fields and tapped the virtues of the folk—and yet all three were exceptionally deliberate artists. Another quality which they have in common is the *sensuousness* of their writing (shown at its most intense in Apuleius' description of Fotis stirring the pot). Psyche's cry to Cupid—'*Mi mellite, mi marite, tuae Psychae dulcis anima*'[2]—has the true Catullan ring, while that insatiable delight in the pageantry of things—in details of colour and shape—which pervades the *Golden Ass* had already cropped up in the *Satyricon*. But neither of these two earlier writers had any 'medieval' quality; Apuleius has several. Just as his frequently rhyming prose anticipates the new versification of the Christian hymn-writers, so his mind is moving away from Graeco-Roman rationalism to a world of miracles and paradoxes. His '*Nihil impossibile arbitror*', though uttered in character and with a grain of salt, is only a step away from Tertullian's '*Credo, quia impossibile est*'.[3] Similarly, his liturgical outpourings in Book XI have much in common both in manner and feeling with those of yet another African and father of medievalism, St Augustine. And even the discreetly managed allegory in the story of Cupid and Psyche (which, so far as we know, was, unlike the ass story, Apuleius' own invention) has a whiff of the *Roman de la Rose*; notice Venus' bitterness towards her enemy Sobrietas and notice especially the birth of Voluptas with which the story concludes. The mystical and romantic attitude to Spring (in Book XI) reads also like an anachronism—as does the one Latin poem which shows this same feeling, the *Pervigilium Veneris*; while the descriptions of processions and spectacles can be closely paralleled by chroniclers' accounts of late medieval *tableaux vivants*. Even in style, in his passion for precise and intricate detail, for the clear line and the clean colour, Apuleius is less like an ancient writer than a medieval painter.

He had then a foot in both worlds, as was natural in an age when rival mystery religions were fighting for man's allegiance while traditional paganism and the rationalistic philosophies were alike out of the running. But his *human* interests were never swamped by his mysticism. The characters in the *Golden Ass*, though mostly not more than vignettes, are definitely flesh and blood. His Venus, like Homer's goddesses, is a jealous and touchy

[2] Chap. 22, p. 107; *The Golden Ass*, trans. W. Adlington, rev. S. Gaselee (1915), v, end 6, '[calling him] her spouse, her sweetheart, her joy, her own very soul'. The more accurate Gaselee revision is used for translation here and below.

[3] Chap. 5, end; *The Golden Ass*, i. 20, '[I think [believe] nothing impossible]'; Tertullian, *De Carne Christi* v. 25–6, more accurately '*Certum est, quia impossibile*' or '[The incarnation] is certain, [just] because it is impossible'.

woman, and Psyche (though her name means Soul and though from one angle her adventures are a quasi-Platonic myth) remains an almost too human girl, an amplified version of the Third Sister of folk story, a charming but naïve Cinderella. And Apuleius, like Homer in the *Odyssey*, not only takes a boyish delight in the strange and unexpected for its own sake but makes his hero react to his fortunes and misfortunes as a credible, consistent and sympathetic individual. In his whole Goyaesque parade of witches, murderesses, adulteresses, robbers, quacks, up-and-comings, down-and-outs, cooks, bakers, and housemaids, there is not one cardboard figure; unlike the stooges in the Hellenistic novels, you can walk round these people and find they have solid backs. More remarkable than that: he has managed to turn a man into an ass and to keep this two-in-one character always acceptable; he never forgets it is a *man* turned into an *ass*. Walter Pater makes here a fitting comparison: 'There is an unmistakably real feeling for asses, with bold touches like Swift's, and a genuine animal breadth.'[4] Apuleius resembles Swift in other things—not only in the wealth of circumstantial detail with which he puts over his fantasies but in the unblinking, sombre strength of such pictures as that in Book IX of the slaves and horses in the mill. While hard enough not to be sentimental, he is not 'hard-boiled'. With typical irony he can put the exquisite story of Psyche into the mouth of a *delira et temulenta anicula*,[5] but there is no irony whatsoever in the moving invocation to Isis at the beginning of Book XI or in the moral drawn by the priest—*curiositatis improsperae sinistrum praemium reportasti*.[6] As the preceding ten books have been a field-day for idle 'curiosity', a modern reader might infer that Apuleius was a 'split personality'. To do so would show no sense of history; one might as well infer that the author of a slapstick miracle play could not have been a good Christian.

Most of the qualities already discussed are excellently conveyed by Adlington's translation; what is inevitably lost is the style of the original. Fronto, tutor to Marcus Aurelius, had been preaching

[4] Pater, *Marius the Epicurean*, rev. (1892), chap. 5, para. 8.
[5] End chap. 22, 132; *The Golden Ass*, vi. 25, 'trifling and drunken old woman'.
[6] Beg. chap. 47, 228; *The Golden Ass*, xi. 15, '[thou hast had] a sinister reward for thy unprosperous curiositie'.

in those days the *elocutio novella*;[7] Apuleius put this into practice. Fronto, in the desire to rejuvenate the language, had gone back paradoxically to the earlier writers of the Republic (we have modern parallels for this); Apuleius accordingly is, in Pater's words, 'full of the archaisms and felicities in which that generation delighted, quaint terms and images picked fresh from the early dramatists, the lifelike phrases of some lost poet preserved by an old grammarian, racy morsels of the vernacular and studied prettinesses'.[8] He is, for instance, addicted to diminutives which probably were greatly used in ordinary speech whence they flowed into modern Italian. His vocabulary is highly exotic (which Adlington's English does not suggest) and some at least of those words which appear in no other writer were presumably his own coining; his object was to find a surprising and yet an appropriate word, to hit the nail on the head but to hit it with a fine flourish. Thus when the ass has been a long time without food his jaws are described as 'full of cobwebs'—*araneantes*: you would look in vain for this word in the Latin classics, but then after all you would look in vain for another ass like this ass.

Apuleius' sentence construction is as unorthodox as his vocabulary. His sentences are often as long as Cicero's but they are not 'periods'; the Chinese boxes of subordinate clauses, the geometrical architectonic, have gone; in their place is an arithmetical or cumulative technique, a succession of fairly short phrases, roughly equal in length and often rhyming, often without conjunctions, just adding up and adding up. As compared with Cicero the syntax itself is simple, but the total effect of this tessellated prose is an unashamed artificiality. Adlington who flows along easily, using generally far more words, loses the chime and glitter. Much of Apuleius could be printed as verse (though admittedly not 'classical' verse), e.g. Psyche's prayer to Ceres:

[7] Marcus Cornelius Fronto, *The Correspondence*, ed. and trans. C. R. Haines, 2 vols, rev. (1929), i. xxxii: 'But the *novello elocutio* of which he speaks seems rather to mean a fresher, more vivacious diction, and a more individual form of expression: in fact originality of style.'

[8] Pater, op. cit. n. 4 above, chap. 5, para. 2; either MacNeice or his publisher John Lehmann has omitted Pater's word 'curious' before 'felicities'.

per tacita secreta cistarum
et per famulorum tuorum draconum pinnata curricula
et glebae Siculae sulcamina
et currum rapacem
et terram tenacem
et inluminarum Proserpinae nuptiarum demeacula
et luminosarum filiae inventionum remeacula.[9]

Euphuism is the word that at once occurs to a reader, but the style of the *Golden Ass*, unlike that of *Euphues*, is a means as well as an end; it carries the story admirably, fitting all the moods—and there are many. Witness the quick description of deepening night: *cum ecce crepusculum et nox provecta et nox altior et dein concubia altiora et jam nox imtempesta.*[10] This is a verbal glamour that English cannot emulate; the Elizabethans, however, coming nearer to it than anyone since, Adlington's version will probably remain unchallenged. For, while in its own right it is very fine prose, it also keeps fresh the honey and salt of the original. *Mel* and *sal*?[11] Both should be welcome to-day.

Pindar: A New Judgment[1]

Pindar. By Gilbert Norwood. (University of California Press. $2.50.)

Readers of Pindar, from the time of the Alexandrian critics, have found him a puzzler; among English schoolboys he is easily the most unpopular source of 'unseens', while even those who keep up their classics after university age are inclined with Voltaire to shun

[9] *The Golden Ass*, vi. 2, 'by the secrets of thy baskets, by the flying chariots of the dragons thy servants, by the tillage of the ground of Sicily which thou hast invented, by the chariot of the ravishing god [Pluto], by the earth that held thy daughter fast, by the dark descent to the unillumined marriage of Proserpina, by thy diligent inquisition of her and thy bright return'.

[10] Ibid. ii. 25, 'till it was dark, and then night deeper and deeper still, and then midnight'.

[11] 'Honey' and 'salt'—i.e., sweet sounds and sharp scepticism.

[1] *New Statesman and Nation* 31: 795 (18 May 1946), 362.

Des vers que personne n'entend
Et qu'il faut toujours qu'on admire.[2]

A new book on Pindar by a humane scholar is therefore extremely welcome. Professor Norwood's comments, like Pindar's own lines, being *φωνάεντα συνετοῖσιν*[3]— addressed to the intelligent—this is not a book for hack examinees, but can be recommended to any lover of poetry, provided he has some small knowledge of Greek. Professor Norwood begins with a smack at the Textual Criticism fanatics who would degrade the study of Greek and Latin poetry to 'a squalid imitation of the applied sciences'. It is refreshing to find a classical professor stating downright that by 'the law of diminishing returns' textual criticism has had its day and that it is now time for the student to use his imagination. Using his own, he has reappraised Pindar while debunking certain usual appraisements of him, in particular those which assume 'that in the criticism of Pindar ethics must outweigh æsthetics'.

The habit of using Greek and Latin classics as tag-books has too often obscured the architectonic virtues of the originals. Where even Horace's odes have not been credited with structure, it is not surprising that people should think they are getting the quintessence of Pindar by running their tongues round detached phrases like the unforgettable *ὅταν αἴγλα διόσδοτος ἔλθῃ*.[4] But this is not enough—just as it is not enough to assess Spenser as a merely decorative or merely musical poet. Pindar, though doubtless a primitive 'thinker', was a highly conscious artist and certainly intended each single ode to be a unity; the question for us is where does this unity lie. Professor Norwood denies emphatically that it lies in any sort of 'philosophy' or propagandist purpose. His chapter on Pindar's 'views' demolishes those many scholars who try to discover in Pindar an ethical system or underlying didacticism and proves, I think conclusively, that the Myth which normally occupies the centre—and the greater part—of each ode is not chosen primarily to point any moral.

[2] Voltaire, Ode 17, st. 1: 'verses that no one understands / and which [nevertheless] it is our duty to admire', quoted by Norwood (1945), 94.

[3] 2nd Olympian 85.

[4] 8th Pythian 96, 'when there comes . . . / A gleam of splendour given of Heaven'—*The Odes of Pindar*, trans. Geoffrey S. Conway (1972), 144.

His analysis of Pindar's muddle-headedness, especially about the relations of *areta*[5] and success, is a tonic antidote to the common assumption of hellenists that all ancient Greeks were distinguished by clarity of thought. Nor does he mince his words about Pindar's narrow-mindedness, pointing out that, though 'patriotic' in a narrow sense, Pindar could not appreciate the then developing patriotism of Athenian democracy because, a Dorian of the Dorians, 'his affection for oligarchy enabled him to identify his country (as we should call it) with his own class'.[6] Similarly in religion Pindar was pious but a naïf, for whom 'gods, as individuals, never right any wrongs that are not directed against themselves: they stand for nothing at all except their own comfort, rank, and privileges, exactly like the Dorian nobles eulogized by Pindar, who were not leaders in any spiritual or moral sense.'[7] But, while thus debunking the conception of Pindar as Sage, he readjusts the balance by rebutting the charge that Pindar, as author of so much commissioned work, had thereby sold out his independence as an artist.

Far more novel than this preliminary placing and displacing of Pindar against his historical background are those chapters (the bulk of the book) which look for his poetic coherence. Professor Norwood concentrates here on the use of mythology and the use of symbols. While mythological allusions are at least half wasted on those unacquainted with their matrix, this long-established device persists among us to-day; Mr Eliot's literary allusions are essentially of the same kind. The important thing is that in Pindar, as in Eliot, the allusions are integral to the poem. While the reference to Eliot is mine, Professor Norwood brings out another point in common between these poets—the observation of καιρός[8] i.e. of the significant highlight, the vivid and pregnant detail, the excerpt from a chain of events which, divorced from its *termini a quo* and *ad quem*, achieves a specious present. He shows how Pindar was innovating here and knew it, herein, though not in his opinions, being far more sophisticated than Homer.

It is however Pindar's symbolism which chiefly occupies Professor Norwood and here he is most illuminating. While

[5] 'Virtue / excellence'. See Norwood, 48 f. [6] Norwood, 68.
[7] Norwood, 57. [8] See Norwood, 168 ff.

disliking the pedantry of the old-fashioned classics don, he does look at the words—but looks at them as living things. Having perceived that Pindar's choice of phrase is often conditioned by the dominant motif of the ode, he searches each ode for a key symbol. Such are the Lyre in the 1st Pythian, the Triple Diadem in the 7th Nemean, and the Pebble in the 10th Olympian (the last a particularly neat example which involves among other pieces of Pindaric sleight-of-hand a play on the word ψᾶφος).[9] This quest for a symbol and its reflections is most fascinating in those odes or passages of odes which have been traditionally dismissed as difficult or irrelevant; thus by some brilliant detective work on the 11th Pythian he discovers a key symbol in the Bee (the Bee being also a name for a prophetess and the ode featuring Cassandra). In the 7th Olympian again, through remembering that Rhodes was the Rose Island, he finds that the image of a growing rose pervades the whole poem. The same method, on a lesser scale, enables him to vindicate allegedly meaningless epithets. Whether all his findings can be accepted or not, Professor Norwood has certainly done much to acquit Pindar of incoherence—and so to send at least one reader back to him with more hope of finding his way through those notorious but glamorous labyrinths.

The Traditional Aspect of Modern English Poetry[1]

Since the appearance of *The Waste Land* by T. S. Eliot in 1922 a great deal of recent English poetry has appeared to people of conservative tastes to be entirely divorced from tradition. Such people find Eliot and his successors essentially unpoetic—in content, metrics, imagery, diction and sometimes also syntax. Yet when we turn to Eliot's critical writings we find him stressing the necessity of an *historical sense* which 'compels a man to write not merely with his own generation in his bones but with a feeling

[9] 'Pebble, accurate count, vote'. See Norwood, 110–14.

[1] *La cultura nel mondo* (Rome), Dec. 1946, 220–4.

that the whole of the literature of Europe from Homer and within it the whole of the literature of his own country has a simultaneous existence and composes a simultaneous order'.[2] If Eliot then is practising what he preaches, why do these champions of tradition find him so unsympathetic?

The answer is that many who talk about poetic tradition mean thereby nothing other than the poetic tradition of the Nineteenth Century; whereas Eliot (who in an essay on Marvell laments 'the effort to construct a dream world, which alters English poetry so greatly in the nineteenth century')[3] would maintain that this latter 'tradition' was merely an enormous aberration. And *world* tradition, quantitatively at any rate, endorses Eliot's judgment; the bulk of the world's poetry is not, in the Victorian sense, 'Romantic'. Let us see therefore what kind of things the nineteenth century 'traditionalists' object to as unpoetic.

They object to irregular or broken metre:

> I am ill; but your being by me
> Cannot amend me; society is no comfort
> To one not sociable; I am not very sick
> Since I can reason of it.
>
> (*Cymbeline*, IV, 2.)

They object to 'flatness' of matter and diction—also illustrated in the above quotation.

They object to any appearance of cerebration, whether it shows itself in play upon thoughts (and words), in rhetoric, or in hard argument:

> Omission to do what is necessary
> Seals a commission to a blank of danger;
> And danger, like an ague, subtly taints
> Even then when we sit idly in the sun.
>
> (*Troilus and Cressida*, III, 3.)

They object especially to an intellectual expression of love:

> . . . When you do dance, I wish you
> A wave o' the sea, that you might ever do
> Nothing but that; move still, still so,
> And own no other function: each your doing,

[2] 'Tradition and the Individual Talent' (1919), *Selected Essays*, 14.
[3] 'Andrew Marvell' (1921), *Selected Essays*, 301.

So singular in each particular,
Crowns what you are doing in the present deeds,
That all your acts are queens.

(*The Winter's Tale*, IV, 4.)

They object to 'cynicism' in general but especially to cynicism about sex:

The fitchew nor the soiled horse goes to't
With a more riotous appetite.
(*King Lear*, IV, 6.)

They object accordingly to the precise mention of anything that can be called ugly. But at the same time they object to images which make the imagination jump:

And pity, like a naked new-born babe
Striding the blast . . .
(*Macbeth*, I, 7.)

They object to 'difficulty' or 'obscurity' but they equally object to a simplicity which they think unworthy of its subject:

And a man's life's no more than to say 'one'.
(*Hamlet*, V, 2.)

And they object to many other things in 'modern poetry' which have also their precedents in Shakespeare, from whom all the above quotations are taken.

The 'Romantic' answer would be to distinguish the spasmodic 'poetry' in Shakespeare from the mass of alloy necessary in such an impure medium as drama. The modern world has been unduly obsessed by the 'liric' but even to the lyric this answer will not apply. A. E. Housman in a famous lecture on *The Name and Nature of Poetry* (1933) plumped for 'poetry neat or adulterated with so little meaning that nothing except poetic emotion is perceived and matters' and quoted as an example of such divine nonsense Shakespeare's lyric in *Measure for Measure*—'Take, O take those lips away . . . '.[4] But this, of course, not only has absolute meaning as a self-contained lyric but has a very important meaning relative to the play in which it occurs. And if we examine

[4] *Measure for Measure* IV. i. 1, cited in Housman, *The Name and Nature of Poetry* (1933), 41.

Housman's own work we find it is consistently 'adulterated' with meaning; he even makes ample use of rhetoric, irony and point—the special tricks of writers who want to plug their meaning. His poems have indeed far more in common with a calculatory meaning-plugger like Horace than with his own chief example of 'pure poetry', Blake.[5] What he was really expressing in this brilliantly phrased but fallacious lecture was a distaste for a certain kind of meaning, i.e. that which in England was especially exploited by the so-called 'metaphysical' school of the early Seventeenth Century. It is significant that it is just this school which—so far as literary influences go—has for nearly thirty years had most influence on the younger poets of Britain. Thus Eliot in an essay on 'The Metaphysical Poets' (1921) admires that 'mechanism of sensibility which could devour any kind of experience'. 'A thought to Donne', he writes, 'was an experience'.[6] And his own poetry owes its strength to thoughts which were experiences—as does that of Britain's next great innovator, W. H. Auden.

Between the two Great Wars of 1914–1918 and 1939–1945 then, the intellect came back into English poetry, stiffening its backbone, widening its range, and linking the present with the past. But here we should avoid overstatement and oversimplification. The early seventeenth century in England was a period, like the present, of transition, a clear-cut scholastic order giving way to a clear-cut scientific one. A poet like Donne had a foot in both worlds, but both were fairly comprehensible. Whereas the new thought of to-day, especially when correlated with world events, appears to have a trend that moves more and more towards chaos. Freud, for example, in exposing human disorders and at the same time explaining them by a new kind of order beneath them, was only making explicit an aspect of truth which some people had for long recognized instinctively and which, since Baudelaire's time at least, had been showing more and more in the arts. Thus the Surrealists, who acknowledge their debt to Freud, are the successors of the French Symbolists (who knew him not) whose method, in turn, had been often, if spasmodically, anticipated in English poetry and had even been vindicated in critical pronouncements by Coleridge.[7]

[5] Housman, ibid., 40 ff. [6] *Selected Essays*, 287.

[7] A *locus classicus* of Romantic criticism—Coleridge's prefatory note to 'Kubla Khan: Or, A Vision in a Dream' (1816), explaining the mystique of its composition in psychological or dream-vision terms.

To-day therefore, while we find the reappearance in British poetry of that 'tough reasonableness' which Eliot admires in Marvell,[8] we also find in places an unprecedented outburst of unreason—a deliberate surrender to the Unconscious. Those who thus surrender are the logical heirs of the Romantics—but they are also the real anti-traditionalists. The Surrealist poet describes himself as a 'modest registering machine'[9] but the word 'poet' means *maker* and in Europe they have usually known what they were making. What the Surrealists build on, however, is an *aspect* of truth, and is admitted as such—though not as a forced foundation—in the poetry of two such different traditionalists as Eliot and Auden. Thus the latter not only practises deliberate ambiguity (the Symbolist technique of *Suggestion*) but is on occasion suggestive without knowing how he does it. And Eliot in an essay has referred to an experience, which he found, *without knowing why*, extremely significant; this supplied him with an image in his poem, 'The Journey of the Magi'.[10] The point is that these two poets, unlike the Surrealists or groups which derive from them such as the very young English 'Apocalyptics',[11] while listening to the promptings of the Unconscious, do retain conscious control, do remain makers—or shapers.

To come now from content to form (remembering that the two are not strictly separable) Eliot and Auden and their adherents will again be found traditional in their methods. Where they differ from the Old Masters, the differences are nearly always merely of *degree*; e.g. if their transitions in imagery seem bolder, they are not bolder *relatively* to their own world where the association method, as distinct from logic, is now fairly generally understood, as is shown even in the 'cutting' of popular films. Eliot's use of the close-up detail:

> I should have been a pair of ragged claws
> Scuttling across the floors of silent seas.[12]

[8] *Selected Essays*, 293.

[9] Cf. MacNeice's article 'Subject in Modern Poetry' (Dec. 1936) above with n. 48.

[10] *The Use of Poetry and the Use of Criticism* (1933), 148; previously cited by MacNeice in *The Poetry of W. B. Yeats* (1941, 1967), 124.

[11] For 'Apocalyptics' see MacNeice's article 'An Alphabet of Literary Prejudices' (March 1948) below with n. 2.

[12] 'The Love Song of J. Alfred Prufrock', ll. 73–4. Eliot more probably refers to the sideling crab than to the lobster.

(where he means a lobster), is another trick very natural in a world of cinema-goers; but Eliot is only availing himself of his world in the same way that Donne availed himself of his. If his is to be called untraditional, then to mention an aeroplane is untraditional. Similarly, Eliot's habit of literary allusions, the insertion sometimes of tags from older poets is analogous to the long-established practice (now not so fruitful) of alluding to Greek mythology. And Auden's personification of Freudian concepts—e.g. in his sonnet sequence, 'The Quest'—has its parallel in Spenserian allegory (a mode which the Victorian would have thought obsolete). Even Dylan Thomas (who from the frequently random appearance of his content might be classed with the latest Romantics, the anti-traditionalists) is in many of his daring epithets following the road of Marvell's 'green thought in a green shade'.[13] Thomas uses more of such epithets and many of them are more startling but (when they do fit) the difference from Marvell, if any, is a difference of degree.[14] It should be added that Thomas, however his matter may seem at times to approach the Surrealists', remains, when it comes to the *pattern* of his poems, *par excellence* a shaper.

Formal pattern—in the narrower sense of metre, rhythm, rhyme, etc.—is a sphere where the 'modern Poet' is most often attacked in the name of tradition by his enemies. 'Form' for these conservatives seems to mean a mellifluous regularity as found e.g. in Swinburne. But to ask for every line to be mellifluous is like asking for everything mentioned to be 'beautiful' (as indeed some of the conservatives do). Swinburne's form is almost inexpressive and heralds the death of content; whereas the classical notion of form is the maximum of expressiveness. Thus Shakespeare broke up Marlowe's blank verse in order to say more things more subtly, and Eliot has broken up traditional verse forms for exactly the same reason. A remark of Gerard Manley Hopkins, which applies to many things, applies here: 'in everything the more remote the ratio of the parts to one another or the whole the greater the unity if felt at all'.[15] Rhythmical variations are *not*

[13] Marvell's poem 'The Garden', st. 6.

[14] MacNeice's note: 'The daring, non-rational, transposition of epithets goes of course back to Virgil.'

[15] Hopkins, 'Rhythm and the other structural parts of Rhetoric—verse', *The Journals and Papers*, ed. Humphry House and Graham Storey (1959), 283.

the death of rhythm. Eliot is a master of variations and in his poems, some of which look almost like 'free verse', the evidence is that the necessary formal unity is felt. Granted that this is so, Eliot's experiments in rhythm in the *Four Quartets* differ again only in degree from Shakespeare's. But in free verse proper (e.g. the now forgotten 'Imagists') the difference is usually of *kind* because there are no rhythmical values to be felt at all. It is to be noted that in England now few young poets attempt 'free verse'. The only valid reason for breaking up forms is to find new forms—when a man needs them. Most of our younger poets, thanks largely to Eliot himself, find themselves not needing even Eliot's technical freedom (which is as well since 'the more remote the ratio of the parts' in versification the better must be the poet's ear—and Eliot has an exceptionally fine ear). Most of the younger generation have returned to more regular forms while trying to be their masters, not their slaves. The trouble, when the best of them take over form (the As-the-Bird-Sings heresy being now nearly dead), is not merely for form's sake; believers in meaning (though not necessarily in a wholly rational meaning) they try, in the European tradition, to convey it by all the means at their command—and most of these means, on analysis, are traditional.

An Alphabet of Literary Prejudices[1]

Apocalyptics. This group of poets compares badly with their predecessor on Patmos.[2] As with the surrealists, I don't mind

[1] *Windmill* 3: 9 (March 1948), 38–42. Editorial note by Edward Lane (p. 1): 'Miss Pamela Hansford Johnson contributed the first essay to this series ['Literary Prejudices']; she was followed by Mr. G. W. Stonier, who arranged his personal comments into an alphabetical frame, a form which was adopted by Mr. Daniel George, Miss Rose Macaulay, Mr. William Plomer, Mr. James Agate, Mr. Geoffrey Grigson, Mr. Edward Sackville-West, Mr. Nigel Balchin and Mr. Louis MacNeice. . . . [I]n future, we shall not be bound to the twenty-six letter formula. . . . Further writers to this series will include Mr. V. S. Pritchett, Miss V. Sackville-West, Mr. L. P. Hartley, Miss Rebecca West and Mr. Louis Marlow.'

[2] A group of writers styled 'The New Apocalypse' and published in several anthologies in the 1940s—*The New Apocalypse* (1940), ed. J[ames] F. Hendry; *The White Horseman* (1941) and *The Crown and the Sickle* (1945), both ed. Hendry and Henry Treece—borrowed the apocalyptic symbolism of John the divine of Patmos from his biblical book of the Revelation or Apocalypse, but linked it to current surrealism and anarchism against the classicism and realism of Auden and MacNeice.

what poets call themselves but I find it a pity when they're *dull*. A reaction against the so-called 'social consciousness school' (not that it was a school) was inevitable and healthy, but the leaders of this reaction protested too much; taking them at their word one expected something fresh, but what one found was mainly derivative. And Urge is fine but uncanalized Urge gets nowhere. A spokesman of the New Apocalyptics wrote in 1941: 'With the war, we are all forced in a sense to become stoics—to depend on ourselves and the universe, the intermediate social worlds having been largely destroyed.'[3] A sympathetic statement, but on analysis all poetry, even Pope, depends on oneself-and-the-universe. But language cannot be divorced from some sort of social world. No more than one can play tennis without a net[4]—or an opponent.

Book Reviewers. I have been among these off and on, so know their occupational diseases. Their worst habit is assuming they know the questions to which the work reviewed provides the answers. Because of this they often make two opposite but equally unjustified demands, (1) that an author should keep 'developing' (discovering every year a new message, new technique or new self—as if there were all that newness under the sun), (2) that an author should stay put in a pigeon-hole—as a lyric poet or a satirist or a realist or what-have-you.

Chunks of Life. Plot and point, like rhetoric, can easily become abuses. All pattern is artificial and most patterns need smashing up on occasions; we cannot, for all that, get away from artificiality. The writer who despises form must still formalize even in selecting his material. To despise 'form' will not bring him nearer reality but may very easily take him further from it. Between the two great Wars many writers forgot this danger. The short story in particular was degraded into a camera to shoot (in most cases) the proletariat. This was not fair either to the art of fiction or to the proletariat. The chance snap of the surface of a living thing will hardly give you its life, while a thousand such snaps will

Hendry in his introduction to the first anthology, 12, quoted from MacNeice's *Modern Poetry*, 27. See Treece, *How I See Apocalypse* (1946); Arthur Edward Salmon, *Poets of the Apocalypse* (1983).

[3] G. S. Fraser, 'Introduction: Apocalypse in Poetry', *The White Horseman*, 30.

[4] Cf. Robert Frost's comment on fair play in verse as in tennis, quoted by MacNeice in his review 'Frost' (July 1963) below with n. 7.

merely give you a headache. The chunk-of-life may be all right for the journalist to write and the man in the tube to read but for serious writers and readers it is a missed opportunity.

Dark God. As D. H. Lawrence was well slapped down in the 'twenties by Mr Wyndham Lewis,[5] there is no need now to take another slap at one who, in spite of his unfortunate effect on adolescents, was a great writer and a godsend. But we still find new dark-godders cropping up with at least as much conceit as Lawrence but without either his eye or his integrity. In English-speaking races very, very few people are *born* dark-godders; to thrust the Dark God upon oneself after reading (say) Freud is, as Freud would have seen, a perversity. One can easily find a sexual significance in everything but one could also find a mathematical significance in everything—so what? Let us get away from nineteenth century snobbery and admit that, so far as our heritage goes, we are Christians. Christ struck a happy balance between East and West and owes his pre-eminence, as Shakespeare does his, to a blend of common sense and imagination. Lawrence had imagination without common sense—and got away with it—but in most people this divorce will degrade imagination itself. Thus we find Mr Henry Miller writing turgid tatty old-fangled romantic exhibitionist prose, trying so hard to be virile and turning out ham.

Enfants Terribles should also be born and not made. The influence of Rimbaud on modern poets, especially in America, is disastrous. Rimbaud was magnificent but a freak. It is not for people like me—nor most probably for people like you—to self-consciously befreak our own Unconsciouses.

Free Verse had to be tried but now—with rare exceptions—ought to be dropped. Some traditional verse forms are like ladders with rungs every few inches; you get stuck in them or stub your toes on them. But a ladder without any rungs . . . ? In the arts bars can be cross-bars and limitations an asset. Verse is a precision instrument and owes its precision very largely to the many and subtle differences which an ordinary word can acquire from its place in a rhythmical scheme.

Gaelic is a fine language but the Gaelic League in Ireland is barking up the wrong flagpole. Eire, having won her rightful

[5] P. Wyndham Lewis in *The Enemy* 1: 1 (Jan. 1927), previously cited by MacNeice in *Modern Poetry*, 69. Lewis launched a full-scale attack on Lawrence in *Paleface* (1929).

independence, should be able to drop this propaganda which nearly all her citizens see through. When will well-known Irish writers who publish nothing but English stop preaching nothing but Gaelic? Which reminds me: I was told lately in Oslo that Ibsen is now being translated into 'true' Norwegian, i.e. into an artificial amalgam of West Coast Norwegian dialects—the great virtue of which is that the Danes can't understand it.

Hypochondria. Writers have often been ill—physically or spiritually—or thought they were; but they used to write in spite of it. Lately, as with Proust, they wrote because of it. More lately still they have been writing that they can't write because of it, and why they can't because of it and how they can't because of it. But if (as they often say) we are all ill together, couldn't they skip these preliminaries? And just either write or keep quiet?

Inspiration does, I believe, stand for something real; most writers, even the most prosaic, are conscious at times of having something 'given' to them. But on deep analysis even what its author may think sheer hackwork may also be 'given' so that the term becomes too general to use as a criterion. It should certainly not be used to bludgeon any poor man who suffers from knowing what he's doing. Remember that Plato, from whom the Romantics derived their theory of Inspiration, thought of it as a bad thing.[6] The simple reader is welcome to go on thinking that Catullus and Burns were purely 'inspired' by love, but it's time that critics ceased to be so naïve. Hardly any poets in the world's history have written 'as the bird sings'. And birds don't sing for publication; think what bores they are when broadcast.

Jargon, though beaten down in some quarters, is still very much with us; look at any American cultural periodical. (By jargon I do not mean the use of technical but the misuse of technical or the use of pseudo-technical terms.) 'Polemic' has exposed it as a social and political evil. It is also of course an æsthetic evil; I am not sure that the very fine poet W. H. Auden is not guilty of jargon in some of his later poems where he throws in catch-phrases from philosophy, theology, etc., etc.[7] Sometimes

[6] Plato, *Ion* ; *Republic* 398 A–B.

[7] The word 'polemic' as 'jargon' *may* refer to Auden's Phi Beta Kappa poem 'Under Which Lyre' (Harvard, 1946), st. 5: 'And every commencement speech / Be a polemic', in which an allusion to 'warlike' in 'polemic' may not be an etymology available to a common reader with no Greek.

apparently just because he likes them and only too often without any likelihood of his readers gathering their particular application.

Kersey Noes—and honest russet yeas[8]—are still recommended by those who take Wordsworth at his word and preach 'the common language of men'. I needn't go into this fallacy as Coleridge exposed it at the time.[9]

Litteræ Humaniores, Honours, School of, has fertilized our literature and social life in the past and I cannot agree with those who think it now quite sterile. But some of our neo-hellenists have painfully one-track minds; they trot out Ancient Athens (an over simplified version at that) as a panacea for everything. And when they come to literature it's both silly and insulting to tell a modern author to write like Simonides or Sophocles. We should rather be advised to read such classics and then go and do otherwise.[10] And who knows? In the whirligig of time otherwise may prove the same as likewise.

Marxists do not as a rule make helpful literary critics (exception: Christopher Caudwell). Like other doctrinaires, e.g. psychologists, they absurdly oversimplify, confusing condition and cause. That a writer is conditioned by his social background is undeniable: so is a mathematician; but the theorem of Pythagoras is neither proved nor disproved by Pythagoras' bank balance. As literary *historians* Marxists are some use along broad lines, but broad lines, alas, in both history and literature are far less than half the battle; it is the brush strokes that count.

New York Intelligentsia. No more tiresome than other intelligentsias (think of the nambiness of London, the malice of Dublin, the dogmatisms of Paris) but tiresome in a different way. Some Americans show as much puritan fervour in amassing facts as others show in amassing money. But when it comes to spending they're as slow in making use of the former as they're quick in making use of the latter. American thought is as cluttered with jargon (see above) as American life is cluttered with gadgets and ads.

[8] Plain, rustic speech, as Berowne in *Love's Labour's Lost* V. ii. 412–13: 'Henceforth my wooing mind shall be expressed / In russet yeas and honest kersey noes.'

[9] S. T. Coleridge, *Biographia Literaria* (1817), chaps 14–20.

[10] Luke 10: 37: 'Go, and do thou likewise', altered according to G. M. Hopkins: 'The effect of studying masterpieces is to make me admire and do otherwise'—*The Letters to Robert Bridges*, ed. C. C. Abbott (1935, 1955), 291.

Oedipus Complex. This won't explain everything nor even the *Oedipus*.

Poetic Plays nowadays are rarely plays. Verse (see F) can obviously make a play more emotionally precise but the verse should be there for the play, not vice versa. Some recent experiments at the Mercury showed an airy-fairy disregard for the first elements of theatre.[11] A programme note on a playwright's versification won't make his lines speakable, let alone dramatic. And while T. S. Eliot's *Murder in the Cathedral* was a fine work of its kind, the way some younger authors of verse plays drag in religion by hotch or by potch strikes me as rather a racket.

Quotations are too often used either to save thought or to show off. Thus dramatic critics use them, as they use tittly-tattly reminiscences, to avoid criticizing the plays they are supposed to have sat through. In poetry Mr Eliot introduced the habit of abrupt quotation in *The Waste Land*, a work which in this as in other things is to be admired but not imitated. In *belles-lettres* quotation has always flourished; this provides a great temptation for well-read writers who could go beyond *belles-lettres*. See 'Palinurus'; *The Unquiet Grave* makes fascinating reading, but it remains a scrapbook.[12]

Rilke has had great influence on certain young English poets. So has Lorca. But, while English verse can never get a Spanish colour, English writers may (once in a moon invisible) have Rilke's sense of dedication; even so, they will hardly share his make-up. Rilke was an extreme case of something; extreme cases are often the cream of the arts; they can give us a stimulus but, N.B., not a technique. Compare Gerard Manley Hopkins whose very peculiar versification reflected his own very peculiar personality.

Self-deprecation among authors never rings true. Why publish a work at all if you think it's no good? (Unless it's a pot-boiler, in which case say so.)

[11] The Mercury Theatre, Notting Hill Gate, London, was the 'little theatre' of poetic drama during the war and early post-war years. There, E. Martin Browne and his company of the Pilgrim Players presented not only T. S. Eliot's *Murder in the Cathedral* and *The Family Reunion*, but such new experiments in religious verse drama as Norman Nicholson's *The Old Man of the Mountains*, Ronald Duncan's *This Way to the Tomb* with masque and anti-masque, and Anne Ridler's *The Shadow Factory*, a nativity play (all three 1945). See Robert Speaight *et al.*, *Since 1939* (1949), 45–52.

[12] Cyril V. Connolly (1903–74), *The Unquiet Grave: A Word Cycle* by 'Palinurus' (pseud.) (1944).

Tough Fiction is usually soft at the core. See all hardboiled Americans, including Hemingway. This very easy formula seemed new about 1920, but Hollywood has long done its worst with it.

Utilitarianism. Compare M. Politicians and scientists and educationalists—and good simple folk—often treat literature as a means to an end. And so of course it can be but it is always something more. As a utility it is double-edged. It is like food, which is useful, but it's also like life—which is useless. When I say useless I don't mean valueless. The U.S.S.R. and Professor Hogben[13] and others haven't yet learnt that there are two kinds of value.

Vulgarity, Fear of, can be as vulgar as anything. Just as a wing three-quarter who's to score in Rugby football must generally hug the touch-line, so creative literature, which by its nature involves personal feelings, must run the risk of sentimentality. But it's better to be sometimes sentimental, over-coloured, hyperbolical or merely obvious than to play for safety always and get nowhere. Virgil, Shakespeare, Dickens and countless others were thrust into touch in their time.

Writing Down to the presumed masses and writing up to a factitious élite are both pusillanimous activities, for in either case the writer is false to his views and to himself. Yet one and the same man can often write honestly and valuably for a small public at one time and for a large one at another; most people after all have lots of different things to say—some esoteric, some 'popular'. What we should never do is write for any public, real or presumed, which is so alien to ourselves that to meet it we have to lie.

X Squared Minus Y Squared. A flower is never a formula, the concrete cannot be reduced to the abstract. The late Professor Susan Stebbing did a necessary job well when she demolished those physicists who thought they had proved that God is a mathematician.[14] But too many people besides physicists still

[13] Lancelot T. Hogben (1895–1975), popularizer of *Mathematics for the Million* (1936, many editions), published many scientific tracts, pamphlets, primers from the 1920s to the 1960s. MacNeice reported in *I Crossed the Minch*, 95, that Compton Mackenzie found *Mathematics for the Million* 'repellent to him as an individualist'.

[14] L. Susan Stebbing (1885–1943), Professor of Philosophy at Bedford College when MacNeice was teaching there 1936–9, analysed the meretricious reasoning of physicists Sir James Jeans and Sir A[rthur] S[tanley] Eddington in her *Philosophy and the Physicists* (1937), 20 ff., 258 ff. See *Philosophical Studies: Essays in memory of L. Susan Stebbing* (1948).

have a hangover from old-fashioned science and assume, consciously or unconsciously, that the abstract is more real than the concrete. This childish and vicious heresy *must* be kept out of literature.

Yogis of the Tame West should live but not publish. Mr Aldous Huxley's later novels seem to me repulsive. Of course it's high time to remember the soul but, whatever that entity is, it's surely not smug? And, if you're preaching a gospel of love, need you look down your nose all the time?

Zanies as heroes of fiction have by now, I suggest, had their day. Steinbeck's *Of Mice and Men* was the *ne plus ultra*. Perhaps in life there are no such things as real heroes, but would it not be nice to reinvent one? Or, if you novelists draw the line at that, could you not at least make your central characters positive? As some people are in flesh-and-blood—don't look now, but they're facing you.

Eliot and the Adolescent[1]

It is a truism that different generations react differently to the same body of poetry; so does any one reader at different points in his own life. Thus in our time T. S. Eliot has been many things to many and many things also to oneself. To summarize this Protean impact is impossible but a reader's memory by checking certain shifts of emphasis may throw light on the organic unity which persists through these permutations. It is useful, I think, to sidestep the broad disputes about 'classicism' and 'romanticism' or 'personal' and 'impersonal' poetry and turning back to the poems themselves to turn back also to one's earlier self who encountered them. I first read Eliot's poems in 1926 during my last term at school. What they meant to me then is something considerably different from what they mean to me now, yet certain constants remain. Undoubtedly my adolescent self (like most

[1] *T. S. Eliot: A Symposium*, ed. [Thurairajah] Tambimuttu and Richard March (Sept. 1948), 146–51. Among other prose contributors were: Conrad Aiken, E. Martin Browne, John Betjeman, William Empson, Wyndham Lewis, Eugenio Montale, George Seferis; among verse contributors: Auden, George Barker, Lawrence Durrell, Anne Ridler, Edith Sitwell, Spender, Vernon Watkins.

adolescents?) missed a great deal in Eliot. What interests me is to try to remember what I *got* from him.

How did we schoolboys come to read Eliot at all? The answer, I am afraid, is largely snobbery; we had seen reviews proclaiming him a modern of the moderns and we too wanted to be 'modern'. We had already seen, if we did not own, many reproductions of Picasso, Matisse, etc., and we had not yet found any English poet whose daring in form and content would rank him as a 'modern' beside them. So we got hold of Eliot and, though at a first reading he seemed unheard-of heavy going, we sensed straight away that he filled the bill. Having said this, I must add, as my solemn belief, that snobbery on the reader's part does not prevent what he reads from hitting him hard and truly. Otherwise there would be no point in adolescents reading books at all; we are all snobs in our later 'teens, if not afterwards.

The paradox of *my* generation, who were aged about eighteen in 1926, is that while (again like most adolescents?) we were at heart romantics, i.e. anarchic, over-emotional and set on trailing our coats, the date of our birth had deprived us of the stock, i.e. the Nineteenth Century, 'romantic' orientation. A year before I read Eliot my favourite long poem had been *Prometheus Unbound* but this had already cloyed; Shelley's enthusiasms were beginning to seem naïve to a child of the Twentieth Century, even to a child who had only fleeting contacts with its over-industrialized, over-commercialized, over-urbanized, over-standardized, over-specialized nuclei. What we wanted was 'realism' but—so the paradox goes on—we wanted it for romantic reasons. We wanted to play Hamlet in the shadow of the gas-works. And this was the opening we found—or thought we found—in Eliot.

What we should have found in the *Four Quartets*, had it been published then, is a puzzler; youth finds it easier to face the end of the world than its beginning. The volume available to us began with 'Prufrock' and ended with 'The Hollow Men'; the last had a more immediate impact than the first. Your adolescent likes a dream atmosphere and among dreams prefers nightmares. The images, the rhythms and the hypnotic, incantatory repetitions of 'The Hollow Men' were not too alien to anyone brought up on the Bible and on Shakespeare's tragedies and even on the autumnal Victorians. In the same way the pock-marked moon of 'Rhapsody on a Windy Night' fell naturally into place beside Shelley's 'dying lady'.[2] But 'The Love Song of J. Alfred Prufrock'? At a first reading I saw no form in it and, with the exception of the mermaids at the end, got little kick from it. And the opening image shocked but did not illuminate—perhaps because I was used to dominantly sensuous imagery, having read at that time very little of the Seventeenth Century Metaphysicals. Realism, the mention of things topical or sordid, I was prepared for—but Wit (in the older sense of the word)? Especially (in spite of my acquaintance with the Cubists), especially Wit about a sunset? The image a few lines later, of the yellow fog rubbing its back and muzzle on the window-panes, was different. It was little more daring than some images in A. E. Housman—the beeches that '*stain* the wind with leaves' or the evening that '*bleeds* upon the road to Wales'[3] (italics mine)—but freed from Housman's ti-tum-ti-tum framework it seemed to pull much more weight. At the time this release from limitations seemed to me mere release; I probably thought of 'Prufrock' as *vers libre* and it was only unconsciously and insidiously that Eliot's extraordinary rhythmical skill rang its bell in my nerves. After a few readings I knew this poem by heart.

But after a few readings what did this poem mean to me as a whole? Certainly not what it means now. All poems, even the most direct of lyrics, are in a sense dramatic but few adolescents—and not many book-reviewers—see this. 'Prufrock' is a dramatic poem in a more precise sense, yet even here the adolescent tends to bypass its dialectic and resolve it all into a one-way outpouring of self. This is perhaps why we used to read this poem aloud in an over-emotional booming monotone. We identified ourselves with Prufrock, having first (in flat contradiction of the text) identified Prufrock with Hamlet. Prufrock was obviously up against the world and so—like all good adolescents—so were we and, all being ill with the world, it was a most exhilarating makebelieve to stand among the dooryards and the sprinkled streets proclaiming one's *Weltschmerz* and announcing at the top of one's voice with the cynical arrogance of virginal youth:

[2] Shelley's short poem 'The Waning Moon' (1820), l. 1: 'And like a dying lady, lean and pale'.

[3] Housman, 'Tell me not here, it needs not saying', st. 3, in *Last Poems* (1922), and 'The Welsh Marches', st. 2, in *A Shropshire Lad* (1896).

I have known the arms already, known them all—
Arms that are braceleted and white and bare.[4]

As a result of this egotistical (romantic, if you like) approach, Prufrock himself ceased to be a character and became a mere mask for the young reader's ego, a mask which like those on the Greek stage also served as a megaphone. Eliot's delicate balance of satire and sympathy was shattered. For us it was merely a case of Prufrock (meaning Us) versus Society; that Prufrock himself is a product of that society was something we chose to ignore. So when he says, 'Do I dare disturb the universe?'[5] we assumed that he could if he wanted to. It will be seen how much we missed in concentrating on the self-pity and the masochism (two things as natural as breathing to an adolescent). The nostalgia of that startlingly 'modern' variation on 'Oh for the wings of a dove!'[6]—

I should have been a pair of ragged claws
Scuttling across the floors of silent seas.[7]

seemed to suit our mood perfectly, even though what *we* had to scuttle away from was not the soullessness of the City office, of a banal sex-life, of a weary intelligentsia, but merely the soullessness of the Sixth Form Room. Yet this travesty of ours was not a complete travesty. Prufrock does embody self-pity and masochism—his creator's as well as his own, I would say (Eliot was under thirty when he wrote him). And anyhow *lacrimae rerum*[8] are not a monopoly of the mature adult.

Of the other poems in the volume the 'Portrait of a Lady' meant little to me then because I did not know that kind of lady; yet this too was rendered memorable, if not intelligible, by its sheer technical brilliance. Eliot's supple line which could so exactly and without fuss convey the slightest *nuance*, change of mood or variation on his theme, seemed admirable even to someone on whom that main theme was largely lost. In such a poem as 'Preludes' we felt more at home. Before ever hearing of Eliot we schoolboys had tried introducing into our own verses such modern

[4] 'Prufrock', ll. 62–3. [5] Ibid., ll. 45–6. [6] Ps. 55: 6.

[7] 'Prufrock', ll. 73–4. Cf. 'The Traditional Aspect of Modern English Poetry' (Dec. 1946) above, n. 12.

[8] Virgil, *Aeneid* i. 462, 'sunt lacrimae rerum et mentem mortalia tangunt', 'there are tears shed for things [even here] and mortality touches the heart.'

properties as telephone wires and combustion engines and this, whatever our elders thought, was not on the whole an affectation. The adolescent is peculiarly sensible of his physical surroundings. To take myself: I had only occasionally visited great cities—London, Belfast, Liverpool, Birmingham—but the fact of these cities was mysterious, compelling, frightening; it was one of the great inescapables of my world which a poet, I thought, must recognize. But, until I met these poems of Eliot, I had not seen it recognized duly. In 'Preludes' I found not only that 'smell' of a modern city which your first visit establishes as part of your mentality but also the human element below that surface, something which even the young and innocent can guess at—

> The conscience of a blackened street
> Impatient to assume the world.[9]

However sheltered our young lives, however rural our normal surroundings, however pre-Industrial-Revolution our education, we knew in our bones, if not explicitly, that this which Eliot expressed so succinctly and vividly, this was what we were up against.

The Sweeney poems had too much salt in them for our liking and, while we relished what we thought was a smack at the Church of England in 'Mr Eliot's Sunday Morning Service', we resented its vocabulary—polyphiloprogenitive, superfetation, pistillate.[10] In a year's time or so W. H. Auden would be deciding that all poetry to-day must be 'clinical' but my school-friends and myself, lacking any scientific training, still thought of poetry more as effulgence (Dr Johnson's word)[11] than as analysis. Such an effulgence, though a sombre one, we found in those also analytical poems, 'Gerontion' and *The Waste Land*. 'The tiger springs in the new year' or 'I will show you fear in a handful of dust'[12]—these were in a language of emotion that we knew, the same language as 'Brightness falls from the air'[13] or:

[9] 'Preludes', iv. 8–9.

[10] Mock-pedantries at ll. 1, 6, 27, of the 'Sunday Morning Service'.

[11] Samuel Johnson, 'Cowley', para. 59, in *Lives of the English Poets* (1779–81): 'the wide effulgence of a summer noon'; slightly misquoted in MacNeice's *Modern Poetry*, 98: '*the wide effulgence* of a summer morn [MacNeice's italics]'. Cf. Johnson's epigraph to the *Rambler*, no. 7, and his *Dictionary*, s.v.'effulgence'.

[12] 'Gerontion', l. 48; *The Waste Land*, i. 30.

[13] Thomas Nashe, 'Adieu, farewell earth's bliss', l. 17, in *Summer's Last Will and Testament* (1600).

> The Son of Morn in weary Night's decline,
> The lost traveller's dream under the hill.[14]

The Waste Land, needless to say, was the poem in this book which most altered our conception of poetry and, I think one can add, of life. We knew few of its literary allusions and it was not till years later that I even bothered to ascertain, for example, the meaning of 'Poi s'ascose nel foco che gli affina.'[15] The cosmopolitan world upon which *The Waste Land* is based was as unknown to us as its anthropological symbolism. Yet it had such an enormous impact on us that I am almost forced to explain it by some such hypothesis as Jung's archetypal myths (as has, I believe, been done by at least one critic).[16] The cinema technique of quick cutting, of surprise juxtapositions, of spotting the everyday detail and making it significant, this would naturally intrigue the novelty-mad adolescent and should, like even the most experimental films, soon become easy to grasp; but that the total complex of mood-and-meaning remains for me now, for all its enrichment by experience and study, qualitatively the same as it was then, strikes me as astonishing. It is possible that at the age of eighteen we knew, however unconsciously, more about waste lands than most earlier generations did—or than any adolescent ought to know. Possible . . . but what is certain is this: to have painted the Waste Land so precisely that those who had never to their conscious knowledge been there could so fully recognize it at first sight and at every subsequent meeting could find it still as real or more so, was the feat of a great poet.

Experiences with Images[1]

How do I use images? In trying to answer this question (and I shall merely scratch the surface) I find it hard to be honest. There

[14] William Blake, *The Gates of Paradise* (1793, 1818), epilogue, ll. 7–8.

[15] *Purgatorio* xxvi. 148, 'Then he [the love poet Arnaut Daniel] dived back in that fire which refines them', quoted in Italian in *The Waste Land*, v. 427.

[16] Maud Bodkin, *Archetypal Patterns in Poetry*: *Psychological Studies of Imagination* (1934, 1963); *The Quest for Salvation in an Ancient and a Modern Play* (1941).

[1] *Orpheus*, 2 (1949), 124–32.

are two such strong but opposite temptations—to oversimplify, make it all sound neat-and-easy (here comes the Master Craftsman counting his brass tacks) and to make it all sound alarmingly but glamorously mysterious (here comes Inspiration falling off her tripod). Still it is worthwhile sometimes, having made the due qualifications, to have a shot at such questions and not abandon the job to people like book-reviewers who either are non-practitioners in verse or write a kind of verse quite different from one's own. And to talk once in a way about one's own work is, I suppose, a good exercise in humility.

Let us start with the necessary truisms. The word 'image' can be used quite narrowly, restricted to such devices as metaphor and simile, or it can be used so widely as to cover nearly everything that goes into a poem. This wider sense is the more correct but the less use in a discussion. To discuss poetry, as distinct from writing it, we have to assume that a poem is *about* something and that this primary Thing is put over by the parading of secondary things which illustrate it. In reality this distinction between primary and secondary is often, if not always, a rationalization after the event, a distortion. It is almost impossible either to explain in what sense a poem is 'about' something or even to define, at all exactly, what any one poem is about. The presumption that one can put one's finger on these primaries has very, very often led critics astray—hence for example their unwarranted contempt for so much of Latin poetry; the critics who take the Latin poet at his face value and assume that he begins with a set theme and proceeds merely to illustrate it or decorate it with similes or mythological allusions fail to see that these latter are really the main point of the poem or, to put it differently, that the apparent theme is the mere occasion for the birth of something new, something achieved through the redeploying and reshuffling of myths or images or, as often with these Romans, clichés. Every good poem must have both a relation to 'life' and a new life of its own; the two things are not the same but, since the relation between them, while different in different poets, seems always to defy analysis, let us, having registered this central fact, pass on up to the surface where discussion is possible though it must be crude.

The word 'lyric' has always been a terrible red herring. It is taken to connote not only comparative brevity but a sort of emotional parthenogenesis which results in a one-track attitude labelled 'spontaneous' but verging on the imbecile. In fact all lyric poems, though in varying degrees, are *dramatic*—and that in two ways. (1) The voice and mood, though they may pretend to be spontaneous, are yet in even the most 'personal' of poets such as Catullus and Burns a *chosen* voice and mood, set defiantly in opposition to what they must still co-exist with; there may be only one actor on the stage but the Opposition are on their toes in the wings—and crowding the auditorium; your lyric in fact is a monodrama. (2) Even in what is said (apart from the important things unsaid) all poems, though again in varying degrees, contain an internal conflict, cross-talk, backwash, come-back or pay-off. This is often conveyed by sleight-of-hand—the slightest change of tone, a heightening or lowering of diction, a rhythmical shift or a jump of ideas. Hence all poems, as well as and because of being dramatic, are *ironic* (in the old Greek sense of 'dramatic irony'); poet and reader both know, consciously or unconsciously, the *rest* of the truth which lurks between the lines. And finally the lyric, which is thus dramatic and ironic, is also—it should go without saying—from the first and, above all, *symbolic*. Language itself is by its nature a traffic in symbols but these symbols are plastic—an endless annoyance to the scientist but God's own gift to the poet; for the poet, who is always trying to say something new, must take the rough and ready symbol of a general A and mould it to stand for his own particular *a*; that is at his least ambitious— sometimes he will mould it to stand for *b* or even *x*. Which procedure is itself both ironic and dramatic; these three attributes of a poem are inseparable.

This leads me on to yet another truism. No one aspect of poetry—such as diction, rhythm, sentence structure, or the so-called 'content' itself—can be fully assessed in isolation from the other aspects; to give a crude example, a word in a stretch of blank verse is not the same thing as the same word in a triolet. So poetic images, far from existing in their own right, can only be considered as parts, or rather as aspects, of the whole which is the poem. To decide whether an image is apt or not, you must first have sensed what the poet is getting at. Thus Wordsworth's diction, though not, as he claimed, 'the real language of men', *was*, in his better poems, the real language of Wordsworth. In consenting therefore to discuss my own use of imagery, I ought by rights to explain what as a poet I am getting at. But this is not so easy, as at different times I have been getting at different things and as at all times (like all poets?) I have been answering questions I was not fully aware of having asked. But I think that, generally speaking, my basic conception of life being[2] dialectical (in the philosophic, not in the political sense), I have tended to swing to and fro between descriptive or physical images (which are 'correct' so far as they go) and *faute de mieux* metaphysical, mythical or mystical images (which can never go far enough). 'Eternity', wrote Blake (Yeats's favourite quotation), 'is in love with the productions of Time'[3] and I have tried to pay homage to both. But the two being interlinked, the two sets of images approach each other. Why in some poems does the mere mention of, say, a flower carry the reader not only out of the garden but beyond his normal horizon?

Of the endless complexities of imagery an astonishing number are exemplified in Shakespeare. Most poets are far more narrowly conditioned; thus few can take both their reading and their physical sensations in their stride and few are equally at ease with the image that pinpoints in and with the image that ripples out. Shakespeare seems at home with all sorts—with description and incantation, with the cerebral and the sensuous, with the functional and the decorative, with the topical and the time-honoured, with the nonsense hyperbole and the algebraic cipher, with the grass that is green and the red that is rhetoric, with learned allusion and first-hand observation, with straight punch and hook, with one-word nuance and embroidered conceit. Most of his successors have been far more choosy, implicitly in their practice (and explicitly, often, in their theories) condemning at least half of Shakespeare's writing. Through the eighteenth and nineteenth centuries literary fashions kept changing but they were always narrow. To-day we are no doubt equally swayed by fashion but, as luck or T. S. Eliot would have it, to-day's fashion makes a good deal more allowance for human multiplicity. Mr Eliot's essay

[2] MacNeice's note: 'Hence my fairly frequent use of oxymoron, the phrase which concentrates a paradox.'

[3] Blake, 10th of the 'Proverbs of Hell', *The Marriage of Heaven and Hell* (1792). In *The Poetry of W. B. Yeats* (1941, 1967), 102, 144, MacNeice cites Yeats as quoting this proverb of Blake and associates Yeats's lines 'The stallion Eternity / Mounted the mare of Time' (*Words for Music Perhaps*, no. 23) with Blake.

on 'The Metaphysical Poets' (1921),[4] which could well be regarded as the modern poet's manifesto, is something far more than a piece of special pleading for Donne and Co. It is a challenge to poets to return, if they can and if they dare, to Shakespeare's catholic receptivity.

There is still, of course, a bias—or biases—in the air and, however good our intentions, we are not all born versatile. In my own experience, with images as with other aspects of writing, I find I can hardly tap certain reservoirs which other living poets draw on freely (e.g., W. H. Auden on one side and Dylan Thomas on another). Yet we all to-day live in much the same mental climate, a common world without which even the 'private worlds' of surrealists could not subsist for a moment, a felt though not articulated creed which supports our various personal heresies. 'Who but Donne', wrote Dr Johnson, 'would have thought that a good man is a telescope?'[5] Any of us here, since the climate has changed, might think it. For two great taboos have been lifted off the modern poet—the taboo on using his brains and the taboo on so-called 'free association'. But properly to analyse this mental climate is a job for the historian-cum-philosopher; let us merely postulate it here as something which inevitably conditions our poems and the images within them.

There is then the mental climate of our time and there are also of course the great traditional evergreens, the natural processes within and without the human body—birth, death, sex, the seasons, the sea, the desert, the stars, etc., etc. That any of these obvious sources should cease to supply images to poets is unthinkable, but which of them we draw on most is determined partly by the mental climate (see the recent boom in the Desert)[6] and partly by our own peculiar background. We only notice what we want to notice, we only react to what we want to react to, and we only get the images we deserve. While I must confess that at times I have deliberately gone out of my way to search for a fashionable image, far more often an image (which may or may not be fashionable) has, whether arriving 'spontaneously' or

[4] In *Selected Essays*, 281–91, much cited by MacNeice.

[5] Johnson, 'Cowley', para. 78, in *Lives of the English Poets* (1779–81).

[6] 'Boom' in desert imagery from Eliot's *The Waste Land* and 'The Hollow Men' (1922; 1925); also from Auden, 'The Flight into Egypt' in *For the Time Being* (1941–2).

arrived at after a search, been accepted because it squared with or—to put it more truly—realized my own first-hand experience.

Now about this peculiar background: all human upbringings are more like than unlike each other but the peculiarities do have their lasting effect. Thus place and time are important. The guinea pig I am using in this article, myself, was born in 1907 in Belfast and brought up on the northern shore of Belfast Lough, i.e., in a wet, rather sombre countryside where linen mills jostled with primitive rustic cottages and farmyard noises and hooters more or less balanced each other. Thus the factory entered my childhood's mythology long before I could place it in any social picture. As for the sea, it was something I hardly ever went on but there it was always, not visible from our house but registering its presence through foghorns.[7] It was something alien, foreboding, dangerous, and only very rarely blue. But at the same time (since until I was ten I had only once crossed it) it was a symbol of escape. So was the railway which ran a hundred yards below our house but N.B. the noise of the trains—and this goes for the foghorns and the factory hooters also—had a significance apart from what caused that noise; impinging on me before I knew what they meant, i.e. where they came from, these noises had as it were a purely physical meaning which I would find it hard to analyse. It is partly this meaning which I am concerned with in an early autobiographical poem (1926):

> Trains came threading quietly through my dozing childhood,
> Gentle murmurs nosing through a summer quietude,
> Drawing in and out, in and out, their smoky ribbons . . . [8]

Clearly these trains were not primarily intended for transportation, they are homey things, familiars:

> And so we hardly noticed when that metal murmur came.
> But it brought us assurance and comfort all the same
> And in the early night they soothed us to sleep,
> And the chain of the rolling wheels bound us in deep[9]

And here comes the other image I have just mentioned:

[7] MacNeice's note: 'The word 'siren' in my poems up till 1940 usually refers to a foghorn.'

[8] 'Trains in the distance', ll. 1–3: *Collected Poems*, 3.

[9] Ibid., ll. 9–12.

Till all was broken by that menace from the sea,
The steel-bosomed siren calling bitterly.[10]

These things sound trivial but they form an early stratum of experiences which persists in one's work just as it persists in one's dreams. Sea (i.e. the grey Lough fringed with scum and old cans), fields (i.e. the very small, very green hedged fields of Northern Ireland), factories (i.e. those small factories dotted through the agricultural patchwork), and gardens (i.e. my father's medium-sized lush garden with a cemetery beyond the hawthorn hedge), these things, being my earliest acquaintances, remain in *one* sense more real to me than, say, wide open spaces, downs, mountains, high cliffs, or woods. For it is these former images which I am more likely to use 'instinctively'.

But still more important than time or place is the family. My father being a clergyman, his church was a sort of annex to the home—but rather a haunted annex (it was an old church and there were several things in it which frightened me as a child). Which is one reason, I think, though I would also maintain that the sound is melancholy anyhow, why church bells have for me a sinister association, e.g. in my poem 'Sunday Morning' (1933):

But listen, up the road, something gulps, the church spire
Opens its eight bells out, skulls' mouths which will not tire
To tell how there is no music or movement that ensures
Escape from the weekday time. Which deadens and endures.[11]

In this example, however (where I was thinking of the Birmingham suburbs), I have rationalized or twisted my original association which would have suggested rather 'escape from the *Sunday* time', i.e. from that stony, joyless anti-time of the church (my upbringing was puritanical) which had seemed to preclude music and movement and the growth of anything but stalactites and stalagmites.

Not only of course was the church down the road but in our own house we had the Bible. My favourite reading at about the age of eight was the Book of Revelation but, long before that, biblical imagery had been engrained in me. On top of this I had more than my share of old wives' tales from our Roman Catholic

[10] Ibid., ll. 13–14.
[11] 'Sunday Morning' (May 1933), ll. 11–14: *Collected Poems*, 23 (where l. 13 ends 'movement which secures').

cook and others, of Calvinist alarums from our Presbyterian housekeeper, and of nightmares from various causes. I also had certain early contacts with both mental illness and mental deficiency (these latter may explain the *petrifaction* images which appear pretty often in my poems, e.g., in 'Perseus'). I should add that our house was lit by oil lamps (not enough of them) and so was full of shadows. And in general the daily routine was monotonous, there were few other children to play with and I hardly ever went away to stay. These circumstances between them must have supplied me with many images of fear, anxiety, loneliness or monotony (to be used very often quite out of a personal context). They may also explain—by reaction—what I now think an excessive preoccupation in my earlier verse with things dazzling, high-coloured, quick-moving, hedonistic or up-to-date.

To leave this earlier stratum and come to my *conscious* working with images, my first book, *Blind Fireworks* (published 1929), where that stratum is the most uncovered, is in spite of that the most artificial or literary. This is usually so with juvenile verse, since the poet, having read more than he can catch up with, either writes too easily or forces—and either way lies sentimentality. *Blind Fireworks* is full of images such as I have described from my childhood but it is also full of mythological tags, half-digested new ideas and conceits put in for the hell of it. Thus in one poem coal, thought of as a black panther (which might just pass?), must needs have been petrified by Circe and, when this panther (which is also a sphinx) comes to life in the fire-place, it is discovered (a) to have horns (biblical reminiscence?) and (b) to be 'sprung in the stirrups of life';[12] this is too much. As for mythology, I was in those days only too happy mixing up Greek, Biblical and Nordic; in one poem called 'Twilight of the Gods' a factory chimney 'skewering the consciousness out of mental distance' is equated first with a totem pole throwing a shadow across Enna, the meadow where Persephone was raped, and then with the tree Ygdrasil, famous in Norse legend; round this the dead gods go dancing while Pythagoras (become a sort of cosmic timekeeper) plays tunes on an abacus until it breaks and the universe has to end—beneath 'the snowflakes of Nirvana'. This sort of thing was

[12] 'Coal and Fire', l. 56: *Blind Fireworks* (March 1929), 47.

not, in the ordinary sense, fake; it sprang from an only too genuine emotion and was indeed poured out 'as the bird sings'—but then birds, unlike men, do not have junkshop minds. The fault of such a poem lies not in its feeling but in its technique. For technique begins in the junkshop, in a process—conscious or unconscious—of sorting out. Free association, when it makes literature, isn't as free as it may look.[13]

Apart from this reckless use of mythological allusion, *Blind Fireworks* shows a few examples of good images (usually sensuous) and a number of arty ones—'the cowled and pilgrim moon'—or of ham clichés—'the Proteus of reality'—or of fancy archaisms—I keep comparing things to galleons.[14] The images are rarely structural; they are too often not merely decoration (which is not necessarily bad) but *random* decoration. In my next book, *Poems* (published six years later), I seem to have taken stock of my junkshop—partly just because I was older and partly because I had been reading T. S. Eliot's essays and had so been diverted from anarchism. Some of these new poems are a bit wandering or casual, the 'morals' are sometimes intrusive, and the images are still too often slipped in for their own sake but the whole thing in most cases (I can't keep saying 'so I think') *does keep in key.*

But as these poems are mainly occasional or descriptive (though no poem, remember, can be purely such), their imagery also is mainly occasional or descriptive. For—another autobiographical note—when I wrote these poems I was for the first time (a) working for a salary, (b) living in a large city, (c) married, and (d), in any proper sense, grown up. The result of this was that I found 'significance' in most of what I saw or—perhaps I should say—in my own act of seeing. But just seeing things is a tourist activity; a poet is allowed to be a tourist just as he is allowed to be a journalist—but only so long as it satisfies him. After *The Earth Compels* I tired of tourism and after *Autumn Journal* I tired of journalism. All poems, as I said above, are dramatic but since

[13] MacNeice's note: 'Almost the most disastrous experience of my childhood is for ever associated in my mind with a doubled-up poplar twig—but I have never yet used this image as a symbol of Evil. Were I to do so, I should certainly elucidate the reference.' See *The Strings are False*, 43.

[14] 'Sunset', st. 3: *Blind Fireworks*, 28; "ΓΝΩΘΙ ΣΕΑΥΤΟΝ", l. 7: ibid., 58; 'Reminiscences of Infancy', l. 4, 'Sunset', st. 1: ibid., 8, 28, *et passim*.

Autumn Journal I have been eschewing the news-reel and attempting a stricter kind of drama which largely depends upon structure. On analysis (though I have never thought of it this way when in the act of writing) this structural tightening-up seems to involve four things: (1) the selection of—or perhaps the being selected by—a single theme which itself is a strong symbol, (2) a rhythmical pattern which holds that theme together, (3) syntax (a more careful ordering of sentences, especially in relation to the verse pattern), and (4) a more structural use of imagery.

In *Poems* and *The Earth Compels* my images, attached to details, were usually details themselves. They could therefore be judged by their *correspondence* to particular objects or events. So when I wrote of a child at a circus (to take an image which some found false) that his face in his excitement 'pops like ginger beer',[15] I was merely conflating two ordinary usages—'popping eyes' and 'bubbling with excitement' (champagne, however, would have been out of key). And in 'The Individualist Speaks' I was safely in the tradition of rhetorical or cerebral image when I wrote (of the modern intelligentsia):

> As chestnut candles turn to conkers, so we
> Knock our brains together extravagantly
> Instead of planting them to make more trees . . . [16]

But when we come to the more important structural type of image the criterion of mere correspondence will often fail us. In this category the *rational* metaphor (like that in the last quotation but here used to support a whole poem) should normally get over to a rational reader (of this kind are the images in my poem 'Convoy'), but those which carry the weight of dream or of too direct an experience will require from the reader something more—or less—than reason. Thus an early poem of mine, 'Snow', has often puzzled people though it means exactly what it says; the images here are not voices off, they are bang centre stage, for this is the direct record of a direct experience, the realization of a very obvious fact, that one thing is different from another—a fact which everyone knows but few people perhaps have had it brought home to them in this particular way, i.e. through the sudden violent perception of snow and roses juxtaposed. My

[15] 'Circus' iii (1937), st. 4: *Collected Poems*.
[16] St. 2: *Collected Poems*, 22.

poem, 'The Springboard', on the other hand, though rational in its working out, begins with two irrational premisses—the dream picture of a naked man standing on a springboard in the middle of the air over London and the irrational assumption that it is his duty to throw himself down from there as a sort of ritual sacrifice. This will be lost on those who have no dream logic, as will other poems of mine such as 'The Dowser' and 'Order to View' which are a blend of rational allegory and dream suggestiveness.

In some poems of this type I have used a set of basic images which crossfade into each other; thus in 'Homage to Clichés' (1935) clichés (by which I meant the ordinary more pleasant sense-data of the sensual man) are imaged by fish coming in to a net, by the indolent self-contained behaviour of cats, by the *chiming* of bells (where the bell itself does not move), and by the ordering of drinks ('the same again') at a bar. In opposition to these is the imminent *pealing* of bells (where the bell does move), the tenor bell being equated with a panther (black presumably) about to spring and with a stone Rameses about to move his sceptre and dismiss 'the productions of Time': 'Never is the Bell, Never is the Panther, Never is Rameses.'[17] These equations are reminiscent of *Blind Fireworks* but, I would claim, much more controlled. The same quasi-musical interlinking of images, with variations on contrasted themes, is used in a recent short poem of mine, 'Slow Movement', and, with a more leisurely accumulative effect, in a recent long poem, 'The Stygian Banks'. Such poems are tentative, though unforced, essays in the genre in which Rilke achieved such astonishing exactitude.[18]

The point about the structural image is that its two-or-more-in-oneness pulls a poem together. Hence the value of the play upon words and even of the pun (see the Elizabethans *passim*). To my present taste this sort of economy—the *twist* of an ordinary phrase, the apparently flat statement with a double meaning—is far more exciting than the romantic elaboration of glamour images. Thus the lines that I am especially proud of in my last book are such lines as (of the aftermath of war in England):

[17] 'Homage to Clichés' (Dec. 1935), l. 72: *Collected Poems*, 60.

[18] R. M. Rilke, great modern German lyric poet obsessed with death as a poetic theme (*Duino Elegies* (1923)), as MacNeice was in 'The Stygian Banks' (May 1946).

> The joker that could have been any moment death
> Has been withdrawn, the cards are what they say
> And none is wild . . . [19]

or (of a tart):

> Mascara scrawls a gloss on a torn leaf[20]

(a line which it took me a long time to find). It is of course hard for me to tell whether such conceits 'stick out' too much. The sort of poem I am now trying to write is meant to be all of a piece. For instance 'Hands and Eyes' (the poem from which the last quotation is taken) is an unusually clear-cut example of dramatic structure; it has a dramatic 'build' which almost falls into three acts (Hands—Eyes—Soul) and within each act there is a cryptographic tangle of close-up details and then—in the next verse—a drawing back, a generalization, a *dénouement*. This poem is made up of images but the point of each image lies in the others around it. I would end by repeating that an image is not an end in itself; only the poem is the end, that dramatic unity which must have its downs as well as ups but which, above all, must be self-coherent. A dull image, a halting rhythm, a threadbare piece of diction, which further that end, are far, far better than a sparkling image, a delightful rhythm or a noble piece of diction which impede it.

Poetry, the Public, and the Critic[1]

A critic is by his nature a go-between. In some periods he can fulfil his nature merely by calling from a window, while in others, as in our own, he has great gulfs to cross. Mr Mortimer and Mr Shawe-Taylor have shown in recent articles how awkward these are in the visual arts and in music.[2] What about poetry?

[19] 'Aftermath' (Feb. 1946), ll. 6–8: *Collected Poems*, 219.
[20] 'Hands and Eyes' (May–June 1946), st. 3: *Collected Poems*, 236.

---

[1] *New Statesman and Nation* 38: 970 (8 Oct. 1949), 380–1. See p. x above.
[2] Raymond Mortimer, 'Artist, Critic, Public', ibid., 37: 941 (19 March 1949), 273; Desmond Shawe-Taylor, 'Thoughts on the Present Discords' (on music), ibid., 37: 954 (18 June 1949), 637. The series was completed by J. M. Richards, 'Architect, Critic and Public', ibid., 38: 982 (31 Dec. 1949), 776–7.

It is no good denying that there is a gulf here, too, but I submit that poetry is a special case. A special case because it consists of words, and we all not only use words but use them *poetically*. Ordinary conversation is nearer to lyrical poetry than to cold prose. It is not only that, when we are conversing, we make little effort to be 'objective'; it is also that the casual remark is unique in the way that a lyric is unique. This is true of even the most banal remark; for the spoken word has, without any effort, that plasticity which the lyric aims at, and which prose (I mean prosy prose, not Elizabeth Bowen or William Sansom)[3] ignores.

Having posited this too often forgotten resemblance between poet and talker, we must, of course, qualify it. Where the poet (at any rate, the modern poet) works by words alone, the talker usually deals more in sounds than in words, while he also can draw upon facial expression, gestures, etc. Moreover, the talker has his listener *with* him, they share their context (they are both in the 4.30 on a wet day when war has been declared), everything is given, the premisses need not be stated. Now the poet, too, does not wish to waste time over premisses, but he is not in the 4.30 facing his listener, and he cannot say something like 'I missed the 4.15 by the skin of my teeth,' and be sure that it will ring a bell; nor can he use those innumerable variations of inflection and timing which turn insignificant words into significant speech. So he must compensate. And, as so often happens in so many spheres, the very effort of making up his deficiencies puts him on a higher plane than that on which he found himself deficient. He reaps enormous interest from his own limitations.

The more difficult a game, the greater the technique required of the players. Technique in itself is not valuable, but more often than not it leads to new values, a revelation, a freedom. Poetry is a very difficult game, but its difficulty has usually troubled the players more than the spectators. This is no longer so—and here we come to the crux of this article. Few of the spectators of any sport know all of its finer points, but most of them do get the hang of what is going on; there is a certain broad pattern which cannot be missed, and they are willing to lump the subtleties. Most of the world's poetry shows an equally broad pattern, though

[3] Elizabeth Bowen (1899–1973) and William Sansom (1912–76), writers of prose fiction, imaginative writing.

it often was not what really interested the poets. To-day the broad pattern is frequently missing. But why should a poet dispense with a broad pattern, and, if he does, how can the reader respond to him?

The pattern of a poem involves both form and content; the content must 'make sense', and the shape must be recognizable as shape. But make sense to whom? Recognizable by whom? There are some poems—and they may be very good ones—whose pattern is too subtle for any but a few to grasp. There are other alleged poems which have neither sense nor shape (in any legitimate meaning of those words), and which, therefore, just are not poems. But to decide what is a legitimate meaning of either word needs very delicate analysis. Every poet knows that poetic sense is not the same thing as common sense or logical sense. And every poet knows that poetic shape is not a mere matter of mechanical symmetry. But does the reader know this? He does—when he himself is poeticizing, i.e. talking; he will then use all sorts of hidden allusions, double meanings, irony, hyperboles, and fancy variations from baby-talk to 'meaningless' swear-words; he will also play many tricks with rhythm. But, once he starts reading, he wants to be spoon-fed; he demands that poetry should be simpler than talk. And the reason for this is patent: the poet is not in the 4.30. It is a reason, however, that has been exaggerated; the poet is not in the 4.30 but he *is* somewhere round in the community. Given a community, however loose a one, there should not be an unbridgeable gulf between its poets and its readers. Gulf there is—in spite of the fact that broad patterns are coming back—but is it unbridgeable?

It should not be, if the 'average reader' were educated. But he will not be *poetically* educated until we have responsible critics. As it is, most critics pass by on the other side. But then most critics, unlike creative artists, are snobs and will only preach to the converted. This is true not only of the æsthetic critic who prefers to leave the 'average reader' on the other side of a gulf, but also of the Marxist critic who says that nothing can be done about the gulf until there has been a revolution. In fact, the Marxist attribution of the gulf to social and economic causes is true, but only partly true—which means it is partly false. There are some poets who are 'difficult'—and therefore anathema to the 'average reader'—because, feeling out of touch with the contemporary world, they have deliberately turned their backs on it. There are other poets who are 'difficult' (e.g., W. H. Auden) because they are only too interested in the contemporary world and are trying to assimilate very difficult and complex bits of it (that world, after all, includes Freud just as Shakespeare's included Machiavelli). So much for difficulty of content. 'Difficulty' of form can similarly be referred sometimes to 'escapism', but more often to the sheer fact, which has little to do with social or economic causes, that the English language has been in use a very long time, and that a vast amount of poetry has been written in it already. A poet's words must at all costs come fresh.

While we need not agree with Mr Eliot that 'poets in our civilization, as it exists at present, must be *difficult*',[4] we must concede that a good deal of poetry to-day *will* be difficult, whether we like it or not. But here the critic can help—not that he often does. While it is not for the poet to keep stating his premisses, few poets to-day can assume, as Homer and even Shakespeare could assume, that his public already knows them. And, since it is no good crying out for a tribal uniformity when we are no longer a tribe, it is the critic's job to point out to the public what angle a poet is writing from and what he is trying to do. But the big critic, who writes whole books, is often plugging some perverse general theory of poetry which leaves no room for seven poets out of ten. The little critic, who writes book reviews, seems compelled, partly by lack of space, partly by laziness, to prefer the snap generalization, the ready-made label, to any decent down-to-earth analysis.

Twenty years ago Professor I. A. Richards exposed the pitfalls that trapped his Cambridge undergraduates when they passed judgments on poetry.[5] Those pits are as wide as ever—'stock responses', the obsession with 'message', etc.—and even professional critics fall easily, not to say happily, into them. In reviewing poetry myself I have fallen into such pits; I have also consciously yielded to various temptations. This has proved to me how hard it is to be either an efficient or an honest go-between. I have had it proved again by reviews of my own poems. According to my reviewers, taken collectively (and I am confining

[4] T. S. Eliot, 'The Metaphysical Poets' (1921), *Selected Essays*, 248.
[5] Richards, *Practical Criticism: A Study of Literary Judgment* (1929).

myself to more or less favourable reviews), I am a writer they can place quite simply: I am a surprisingly feminine, essentially masculine poet, whose gift is primarily lyrical and basically satirical, swayed by and immune to politics, with and without a religious sense, and I am technically slapdash and technically meticulous, with a predilection for flat and halting and lilting Swinburnian rhythms, and I have a personal and impersonal approach, with a remarkably wide and consistently narrow range, and I have developed a good deal and I have not developed at all. Most living poets have been similarly treated by reviewers. Can something be wrong somewhere?

Of course something is wrong; the critics may or may not have taste (and, as Mr Eliot has said, 'genuine taste is always imperfect taste'),[6] but very few of them have any critical standards. And they will not acquire such standards till they remember that poetry is made with *words*. For the critic of poetry has one great advantage over the critics of music and painting: words are not notes of music or blobs of pigment, and they could not (*pace* some lunatics) be abstract even if they tried (though they do, of course, have a sound or colour which is distinct from their dictionary meaning). There is therefore some relation between words as used by a poet and words as used by the public, which includes the critic. There is also a relation between verse rhythm and speech rhythm. The poet introduces new subtleties into both rhythm and meaning (the two interact on each other), but he is still only using with a new precision tools which lie ready to everyone's hand. And he is using them for an end which is as much social as personal; as Christopher Caudwell wrote, 'the instinctive ego of art is the common man into which we retire to establish contact with our fellows.'[7] The critic then must start by remembering that words are a means of communication. What is the poet trying to communicate? And does he bring it off? To answer these questions may involve construing the prose 'meaning' of the poem or analysing its prosody; it will certainly involve hard work. When the critic has thus discussed what kind of poem he is dealing with, and whether it is good of his kind, he can then,

[6] Eliot, *The Use of Poetry and the Use of Criticism*, 35.

[7] 'Caudwell' (pseud. of Christopher St John Sprigg), *Illusion and Reality: A Study of the Sources of Poetry* (1937), 155.

if he likes, go on to a third and more difficult question: what is the value of this kind of poem?

But a word of warning. Many poets to-day, whatever they may call themselves, are anarchists.[8] There is a place for such, but there is not much place for the anarchist critic. Here I agree with Mr Eliot who, in 1923, stood out against Mr Middleton Murry's championship of the Inner Voice.[9] Anarchists should sing but they should not speak. 'Criticism's central job', as Mr E. M. Forster said lately, 'seems to be education through precision.'[10] The 'average reader' (who is not the same person as the 'average man') is far more educable that intellectuals think. But neither salon shibboleths nor Inner Voice gush will help him. A poetry reviewer would be more use if he saved some space from generalization and prejudice to discuss such preciser points as the relation between a poet's sentence structure and verse pattern, or his use of the curtain line or his variation of stresses. He must indicate to the reader where to look for the *order* in the poem. But, above all, though he can only do this approximately, he must try to suggest the reason for this order: what is the poet getting at? For it may very well be something that concerns the public, too.

A Poet's Play[1]

The Cocktail Party. By T. S. Eliot. (Faber, 10*s*. 6*d*.)

Poets in our time who wish to be playwrights only too often make their primary aim the writing of plays *in verse*. This is very much putting the cart before the horse. Verse dialogue has certain great advantages over prose dialogue, but the dialogue itself is more than its raiment or, to put it differently, the verse must be dramatic first and verse second. It is a great pity that poets should be blind to this since potentially they are better equipped for dramatic

[8] i.e. the 'Apocalyptics': see the opening para. of MacNeice's article 'An Alphabet of Literary Prejudices' (March 1948) above with n. 2.

[9] Eliot, 'The Function of Criticism' (1923), sects ii–iii, *Selected Essays*, 15–18.

[10] Forster, 'The *Raison d'Être* of Criticism' (1947), *Two Cheers for Democracy* (1951), in The Abinger Edition of E. M. Forster, ed. Oliver Stallybrass, xi (1972), 107.

[1] *Observer*, no. 8292 (7 May 1950), 7.

writing than the ordinary prose-writer. This is because a short poem, as compared with a novel or short story or essay, not only contains or resolves a far more concentrated conflict but is also far less static, consisting as it does of *direct* speech (either monologue or concealed dialogue). Where the lyric has affinities with the play is not in its 'poetic' frills of diction, imagery, etc., but in its total form (which is one of very carefully balanced stresses) and in its omission of padding. Thus, in one very important sense, even Mr Rattigan's plays are more poetic[2] than the most 'poetic' kind of prose—say *Marius the Epicurean*.[3]

Mr Eliot's great virtue as a playwright is that he has progressively been recognizing this. If *Murder in the Cathedral* was still too static, too much of a charade and overburdened with rhetoric, *The Family Reunion* marked a great advance—which seemed to some of its author's admirers a retreat. In *The Cocktail Party* he has advanced—or retreated—further; more coherence of the parts, less rhetorical padding, no chorus, no playing to a gallery either highbrow or High Church. Mr Eliot has attempted here something very daring and well worth doing. He has taken the ordinary West End drawing-room comedy convention— understatement, upper-class accents and all—and used it as a vehicle for utterly serious ideas, for what he has elsewhere called 'the world around the corner'.[4] To make that world convincing you must make your corners convincing and we all can accept those of a London barrister's flat. This play has proved that neither an extra dose of 'realism' nor an extra dose of comedy need mean short rations for the spirit.

Where *The Cocktail Party* fails, the failure is due not to its chosen limitations but to certain deficiencies or certain wrong emphases which could have been adjusted within those limitations. Many critics have already found fault on grounds of *realism* with the 'election' of Celia, but this is the wrong criticism of her sudden flowering into sainthood. Such metamorphoses do

[2] Sir Terence Rattigan (1911–77), author of such prose plays as *French Without Tears* 1936 and *The Winslow Boy* 1946.

[3] Walter Pater's philosophic romance of a young Roman pilgrim-soul (1885).

[4] Probably from Chorus 1 from *The Rock* (1934): 'The desert is not only around the corner, / The desert is squeezed in the tube-train next to you, / The desert is in the heart of your brother'—The Rock speaks, st. 2.

take place and with as little warning in real life, but the arts, as Aristotle knew, must often give more warning than life does.[5] Celia's potentialities should have been more strongly 'planted', as they could have been even in a cocktail party setting. But the transfiguration of Celia, like other though minor flaws in the Third Act, does not invalidate Mr Eliot's attempt to beat the commercial playwright on his own ground—or rather to stake out that ground for serious drama. This is not the only ground for serious drama, but poets would do well to remember it if their only alternative is to fly off, as T. E. Hulme put it, 'into the eternal gases'.[6]

Great Riches[1]

Collected Poems of W. B. Yeats. (Macmillan. 15*s*.)

Why we should have had to wait over ten years for a complete edition of Yeats's poems is hard to imagine; still, here and most welcome it is. These five-hundred-odd pages show this most paradoxical poet chopping and changing through half a century, standing his earlier notions on their heads, loosening or tightening his rhythms, æsthetically—and ascetically—fey in his youth, rumbustious and hardheaded in his old age, symbolist and rhetorician, mystagogue and gossip, near-fascist and anarchist, lutanist and trumpeter by turns, yet always remaining himself.

In the first poem in the collection comes the famous line 'Words alone are certain good,' and Yeats remained to his death first and foremost a lover of words; but his love for other things widened enormously, and the later Yeats would have renounced that other sentiment in the same poem—'Then nowise worship dusty deeds.'[2]

He remained throughout his life an extremely artificial writer, yet his poems at their best feel inevitable—which is a poem's way

[5] Aristotle, *Poetics*, 15, 1454ª 33–6, on the probable or necessary outcome of character as of action—a unity of character just as there must be a unity of action.

[6] T. E. Hulme, 'Romanticism and Classicism', *Speculations* (1924, corr. 1936), 120: 'You might say . . . that the whole of the romantic attitude seems to crystallise in verse round metaphors of flight. [Victor] Hugo is always flying, flying over abysses, flying up into the eternal gases.'

[1] *Observer*, no. 8308 (27 Aug. 1950), 7.

[2] 'The Song of the Happy Shepherd' (1885), ll. 10, 22.

of being natural. He said of himself that as he grew older his poetry grew younger, and his latter-day text seems to have been the paradox 'that Hamlet and Lear are gay'.[3] He deplored no less bitterly than some younger poets much that was happening in the world, but, unlike them and partly perhaps because of his cyclic philosophy and his doctrine of the Masks,[4] he was too much of a tragedian ever to become a pessimist.

> What matter? Out of cavern comes a voice
> And all it knows is that one word 'Rejoice!'[5]

Looking again at the whole body of his poems one is reminded with surprise of the very different *kinds* of poetry in which Yeats excelled. And in which, it must be added, he sometimes failed. There is not a poem which I would wish excluded from this book, but there are several which in themselves (i.e. divorced from their fellows and their author) are trivial, sentimental, over-mannered, half-digested, or just, many would judge, plain silly. Thus both the Gaelic mythology of the earlier poems and the would-be rigid symbolism of the Phases of the Moon are sometimes imposed upon a poem rather than fused with it.

Yet, as Orwell might have put it, some of these inferior poems remain good bad Yeats; at any rate they are Yeats and nobody else. He was such a different type of poet from most of his contemporaries that even his failures hold the value of that difference for us. And, while it is salutary that a sharp inquisitor like Mr Robert Graves should pick 'Innisfree' to pieces,[6] the spell of Yeats is so strange, and any spell these days is so rare that, when in doubt about Yeats, we might well, I think, give him the benefit of it.

In any case how remarkable and how varied are his triumphs: 'No Second Troy', 'The Peacock', 'An Irish Airman Foresees his Death', 'Easter 1916', 'The Second Coming', 'Leda and the Swan', 'Byzantium', 'Longlegged Fly'. And many others. Many readers to-day will prefer poems like the above, his more 'athletic' writing (to use his own word), to the feminine or epicene languors of the Celtic Twilight, but it must be conceded that those too are

[3] 'Lapis Lazuli' (1938), st. 2.
[4] See Yeats, *A Vision* (1925; rev. 1937, 1956), 73–80.
[5] 'The Gyres' (1938), st. 2.
[6] Graves, *The Common Asphodel* (1949), 186–8.

good of their kind and that Yeats's astonishing development does show a continuity.

In this edition the earlier poems appear in his revised versions. As these revisions were often misguided, it is to be hoped that some time we shall be granted a complete variorum edition.[7] In the meanwhile let us be grateful for a book which contains great riches; it is also the record of an artist who remained single-minded in a world of trimmers and who, for all his posing, had integrity.

The Faust Myth[1]

The Fortunes of Faust. By E. M. Butler. (Cambridge University Press. 30*s*.)

Modo he may be called and Mahu, but the Prince of Darkness is rarely represented as a gentleman.[2] Nor are the magicians who have truck with him usually gentlemen. Diamond tries to cut diamond, but both the diamonds are false. Professor Butler's new book rams this home as ruthlessly as did the two earlier parts of her trilogy, *The Myth of the Magus* and *Ritual Magic*.[3]

This most painstaking and erudite study of Faust's antecedents, parallels, and aftermath is like a great Hindu temple, massive and encrusted with grotesques. It makes fascinating if bewildering reading, the conclusion of which is a paradox. The practice of magic and the ritual texts emerge on the whole as thoroughly boring, yet Professor Butler makes good her contention that these were the seeds of some of the world's greatest poetry and drama.

Hereby hangs another paradox—the vulgar and trivial character of the historical Faust of whom Professor Butler remarks, 'Only the lowest and the last of mortals would do to represent magic in the sixteenth century'; and yet this was the creature of whom both Marlowe and Goethe, in their very different ways, made a haunting and cosmic hero. Even their Fausts, of course, have

[7] Edited seven years later: see MacNeice's review (Dec. 1958) below.

[1] *Observer*, no. 8393 (13 April 1952), 7. MacNeice translated Goethe's *Faust* into English verse with E. L. Stahl, in an abridged version for radio 1949 (1951, 1965).

[2] 'The prince of darkness is a gentleman. Modo he's called, and Mahu': Edgar in disguise, in Shakespeare, *King Lear* III. iv. 143–4.

[3] Eliza Marian Butler, *The Myth of the Magus* (1948); *Ritual Magic* (1949).

their obvious feet of clay. Marlowe's suffers, like his other hero, Tamburlaine, from what Professor Butler has called 'the discrepancy between the lofty tone of the prayers and conjurations and the puerility of the "secrets".' Who has not felt the bathos of Tamburlaine's 'sweet fruition of an earthly crown'?[4]

Goethe's Faust similarly never quite lives up to his early monologues in Part I. His land-reclamation in Part II, Act V, is admittedly more admirable than the activities of most medieval magicians who hitched their stars to a wagon of stolen treasure, but is it good enough? The answer must be 'No.' Professor Butler's very detailed researches, ranging from the Median mages to Madame Blavatsky, from Akkadian-Chaldean inscriptions to Aleister Crowley,[5] only too fully ensure that the reader, as she puts it, will 'whilst marvelling at the strength of tradition, grieve over the limitation of human desires.'

Yet the Faust myth remains ever exciting, for 'human desires' should no more be taken at their face value than lyric poems should be judged by their titles or their nominal subjects. Earthly crowns, buried treasures, magico-practical jokes, they all can carry a great symbolical value. And the Faust myth itself is so plastic that everyone can mould it to his purpose; most readers of this book will be astonished not only by the number of authors (mainly German) who have tackled this theme, but by the variety of treatments it has received. We all know that Marlowe and Goethe are poles apart, but few will have heard of the puppet play at Augsburg with its wonderful speech by the Devil: 'Ah, Faustus, if there were a ladder stretching from earth to heaven, made of swords instead of rungs so that I should be cut into a thousand pieces with every step I took, yet would I still strive to reach the summit, so that I might behold the face of God but once more, after which I would willingly be damned again to all eternity.'[6]

The Fortunes of Faust ends with Thomas Mann and Valéry, the former's novel,[7] being, according to Professor Butler, 'an elaboration of Kierkegaard's notion that genius is *ipso facto*

[4] Christopher Marlowe, *1 Tamburlaine* (1590) II. vii. 29.

[5] Mme Helena Petrovna Blavatsky (1831–91) and Aleister Crowley (1875–1947) were quasi-mystical authors and founders of secret societies; both were well known to W. B. Yeats.

[6] Quoted in Butler, *The Fortunes of Faust* (1952), 103.

[7] Thomas Mann (1875–1955), *Doktor Faustus* (1947).

sinful'. Mann's version of the story, which significantly goes back beyond Goethe to the early Faustbook of 1587, is something new but the fortunes of Faust do not end there or with Valéry. Professor Butler stresses that there is a clash in Faustian literature between the optimism of ritual magic and the 'dark commentary' made upon that same by Christian mythology.

Literature thrives on such clashes and the world to-day is full of them. Frustration and optimism combined are the source of all our Utopias and panaceas—Marxist, psycho-analytic or romantic-anarchist. Each of these could throw up a Faust in literature and to each such Faust we should add a Casper, the satyr-play to the tragedy, the clown who cuts through the knots.[8] But whether like Goethe we save any particular Faust or whether like Marlowe we damn him is entirely up to us—for both these solutions can be right.

A Reading of George Herbert. By Rosemond Tuve. (Faber, 25*s.*)[1]

Between the two World Wars, when poets still thought it was proper to learn and to think, the 'Metaphysicals' came palpably back into fashion. Mr Eliot's essay on the subject in 1921, elicited by Professor Grierson's excellent anthology, *Metaphysical Poetry from Donne to Butler*, provided at least two generations of British intellectuals with a set of aggressive quotations: 'something had happened to the mind of England' in the latter half of the seventeenth century; 'a thought to Donne was an experience', whereas after his time 'a dissociation of sensibility set in', the amending of which, it was implied, had been left to ourselves, it being up to us to 'form new wholes' from falling in love and reading Spinoza, from the noise of the typewriter and the smell of cooking.[2] From this parallel between Donne's time and our own the conclusion was drawn 'that it appears likely that poets

[8] Casper—name of the farcical double of Faust in popular puppet plays.

[1] *New Statesman and Nation* 44: 1123 (13 Sept. 1952), 293–4; published under the title 'Books in General'.

[2] 'The Metaphysical Poets', *Selected Essays*, 287.

in our civilization, as it exists at present, must be *difficult*',[3] a conclusion most gratefully accepted by Mr Eliot's juniors to most of whom Chaucer and Wordsworth were merely bores and Homer—if they had analysed it—excluded by definition from the realm of poetry. It is possible that Mr Eliot overstated his case in this memorable essay; it is certain that his disciples overstated Mr Eliot and fell, in his name, into a narrow view of poetry as such and into a distorted view of the Metaphysicals, neither of which views he would have countenanced. Thus these younger men assumed that the *prime* virtue of the Metaphysicals was the audacity of their conceits, concordantly with that, they spoke of 'the Metaphysicals' but what they usually meant was Donne.

Donne is, of course, a young man's poet and even his religious poems were acceptable to an irreligious youth because of his violent or morbid elements, the persistence of the Old Jack. It is regrettable therefore, but hardly surprising, that during the Twenties and Thirties so little attention was paid to George Herbert. Here was a poet who appeared never to have been unregenerate and whose poetry was *all* religious—and Church of England at that; true to the traditions of that Church, it lacked both the rhapsodic flights of a Catholic mystic, such as Crashaw, and the brute drama of the Puritans. Certainly, in the twentieth century the Church of England does appear often to be a *via media* in the worst sense of the term, a bad Samaritan's middle of the road, a middle that knows neither Heaven nor Hell. But it is crude history to assume that it was always such and even today one can sometimes stumble on a member of that body, even on a parson, who not only practises the Christian ethic (which of itself would not make him a Christian) but feels and lives in the Christian legend, waking up elated on Easter Day and even, like Herbert, regarding the fifty-one other Sundays as 'days of mirth', as pillars 'on which heav'n's palace archèd lies'.[4] Such people are very rare and almost never poets. Herbert who, with all allowance made for the hagiographical slant of Walton's *Life*, was to all intents and purposes a saint, was also a poet—and a fine one. Sheer curiosity, one would have thought, should make us read him.

[3] Ibid., 289. [4] Herbert, 'Sunday', st. 4.

On purely technical grounds it is a pity that young poets looking for models to the early seventeenth century should have so ignored Herbert and pondered so much on Donne; Donne is, I think, the greater poet but he is certainly a worse model, just as Gerard Manley Hopkins is a worse model than Yeats. To write like either Donne or Hopkins you have to *start* peculiar, whereas, while both Herbert and Yeats are also peculiar in their very different ways (for few of us are bound either for the Kingdom of God or for the Celtic Twilight), their peculiarities are not so violent or deep as to subordinate their work to their psychology. To put it differently, Herbert and Yeats are more *classical.* Dr Johnson, discussing the Metaphysical poets in his life of Cowley, does not mention Herbert and this is significant, for Johnson there is doing unkindly what we in our time may have done in mistaken adulation, picking out strained conceits which throw their poems out of balance—and he picks them out because they *stick out*. Herbert's conceits very rarely stick out offendingly but are usually organic parts of a whole; nearly all his poems are in balance. His first virtue is *construction*; equal second are the sureness of his music and the purity of his diction which, of course, are part and parcel of construction. He has his pedestrian lapses—'Most things move th'under-jaw; the Crocodile not'[5]—but, compared with most English poets, he maintains a very high standard while, compared with most religious poets, he remains masculine, crisp.

If poets can be forgiven for misreading earlier poets (since even misreadings can be fertile), the non-writing reader should not be encouraged to do so, certainly not by professional literary critics. Yet this is just what some critics are doing today, which is why it is most refreshing to find Professor Rosemond Tuve opening a withering, yet not unfriendly, fire on them. In her own words, she is using 'tools which the newer criticism would like to relinquish' in her allegiance to the old-fashioned 'idea that a poem is most beautiful and most meaningful to us when it is read in terms of the tradition which gave it birth'.[6] She opposes the new heresy 'that understanding is not valuably or reliably advanced by information *directly* bearing on *consciously* intended meaning' (italics mine) and deplores 'the wilful modern divorce between

[5] 'Providence', st. 35. [6] Tuve, 98 and 22.

"scholarship" and "criticism"'.[7] Not that she ignores the fact that scholarship can desiccate and kill—she has far too much humanity and humour—but she sees that the opposite school is attended by the same danger: 'much modern criticism does not, in fact, seem to make the poetry more pleasurable . . . The same old modern swindle, process substituted for essence, has caught us again.'[8] How true that is; with all due credit to Freud for his hugely fruitful illuminations, in the spheres of literature and the arts—and in that of personal relationships when two or three are gathered together to decide what is wrong or even what is right with their friends—new analyst only too often is but old bore writ large.

One of Herbert's longer and most striking poems, 'The Sacrifice', was discussed by Mr William Empson in his *Seven Types of Ambiguity*, a discussion which Professor Tuve considers brilliant but 'inadequate'.[9] She accordingly proceeds, for the first half of this book, to correct and fill in the picture, on the assumption that 'much that is "outside" a poem to us was well inside it to our forefathers'.[10] Drawing on a mass of evidence, including stained-glass windows and illustrations to Hour-books (some fascinating samples of which are here reproduced) she establishes many secondary points such as Herbert's 'habit of freely reading New Testment meanings back into Old Testament images' (e.g., his acceptance of 'Moses as a type of Christ which is a commonplace of iconography', the piercing of Christ's side becoming 'almost synonymous with Moses' striking of the rock to bring forth the stream of saving water').[11] She also proves her more general contention, as regards 'The Sacrifice', that 'all the types of ironic contrast, upon which Herbert's poem is constructed and which work down into its details to give it that ambiguity, density, and ambivalence of tone that we think of as so especially "Metaphysical", are explicit in the tradition'.[12] Thus she cites such precedents from medieval lyric as this magnificent (Metaphysical? modernist?) ironic-epigrammatic paradox:

[7] Ibid., 158 and 42. [8] Ibid., 20.
[9] Empson, *Seven Types of Ambiguity* (1930; 3rd edn 1965), end chap. 7, 226–33; cited by Tuve, 23.
[10] Tuve, 27. [11] Ibid., 29. [12] Ibid., 55.

And in thy soul I set free will,
And now I hang on Calvary hill.[13]

But all this, as she sees after proving that Herbert was not the innovator we thought him, raises the question of 'originality'. I think we must agree with her when she concludes:

Were every element in some of Herbert's images as worn and familiar as a proverb, the cadences and rhythm of his language would yet give them for us the uniqueness of something never before heard. . . . *Tone of voice is a component of meaning.*[14] [Italics mine.]

In fact, and more perhaps in some of the many poems which she does not discuss, Herbert does seem to invent some things for himself (not all his conceits surely are 'traditional'?) just as many of his poems seem to make a pretty full impact on a reader who lacks Professor Tuve's special knowledge. But she has done a great service to lovers of Herbert and of poetry in bridging a pseudo-historical gulf which in no genuine way heightened the stature of the 'Metaphysical' poets who were stranded on the near side of it. When we realize 'that Metaphysical wit and concord of unlikes in an image is precisely the operation, much condensed, of the old (and maligned) allegorical mode of writing',[15] then possibly—however much and however justifiably we may prefer condensation to expansion—we shall even think of re-reading Spenser. For he is, as Professor Tuve says, 'the most notable example of a great poet all but lost to modern readers', lost because they—and readers for centuries 'in a world that expected religion to be science and poetry to be the *literal* truth'—had lost the technique for reading him.[16] Inasmuch as we are dropping those two expectations, some such technique, or rather 'habit of mind', is possibly now coming back to us—and about time too.

Thus many of Mr Auden's poems have an affinity with such allegories in miniature as Herbert's 'Time' or 'The Quip', 'The Bag' or 'The Pilgrimage'; the affinity extends from the habit of mind itself to the modulation of tones, the entrances and exits, the vitalization of clichés, the organization of the stanza. But there

[13] No. 105 in *Religious Lyrics of the XVth Century*, ed. Carleton Brown (1938); cited by Tuve, 78.
[14] Tuve, 80.
[15] Ibid., 134. Quoted again in MacNeice's *Varieties of Parable* (1965), 50.
[16] Tuve, 134–5. See MacNeice's Spenser review (Jan. 1963) below.

is one great difference. Symbols, as used by Herbert (though not by many poets since), being less fluid than other brands of metaphor, imply 'a whole system of traditional and publicly known correspondences' whereas now 'things have come to such a pass that we do not even recognize great numbers of symbols' belonging to our own tradition which is half Judaic-Christian and half Græco-Roman.[17] Herbert after all was not only a saint but a classical scholar and as either was congenial to his public. A modern poet, like Mr Auden, may be highly versed in a number of mythologies, but how many of his readers can be counted on to recognize Elijah or Pentheus,[18] or St Paul or Baldur[19] or Tristan—or even Narcissus?

Yeats's Plays[1]

Collected Plays. By W. B. Yeats. (Macmillan. 18*s*.)

Though Yeats the playwright is not in the same class as Yeats the poet, this master of words who was not a master of dialogue, this theatrical pioneer who knew so little of the theatre, was the cause of drama in others; as even George Moore put it, speaking of the Abbey, 'all the Irish movement rose out of Yeats and returns to Yeats'.[2] And even the many malicious passages in Moore, such as the description of his projected collaboration with Yeats in the writing of *Diarmuid and Grania*,[3] bring out the fact that Yeats, though lacking a dramatist's equipment, did see clearly, long before Mr Eliot, certain crying needs in the theatre which theatre people ignored, in particular a fresh subject-matter and a rediscovery of style.

His earliest play, *The Countess Cathleen*, which he later dismissed as no more 'than a piece of tapestry', has at least a better

[17] Tuve, 104 and 108.

[18] In Greek myth, King of Thebes, denier of Dionysus' divinity, and torn to pieces, as in Euripides' *The Bacchae*.

[19] In Norse myth, god of light, killed through Loki's mischief.

[1] *Observer*, no. 8422 (2 Nov. 1952), 8.

[2] George Moore, *Vale* (1914), *Hail and Farewell*, III, end of chap. viii.

[3] Moore, *Ave* (1911), *Hail and Farewell*, I. xv.

theme and more excitement and verbal vitality than the verse plays of the Victorians; if it is tapestry, it is stronger tapestry than Tennyson's. And the extraordinary little plays allegedly modelled on the Noh plays of Japan, while affording the wits of Dublin—and not without reason—an endless source of irreverent punning, were a worthwhile experiment *if only for what they left out*.

For subjects Yeats went to the early Irish legends and to the living though late Irish peasantry; both sources are by now pretty well exhausted. Loathing the work of Ibsen and his 'leading article sort of poetry' and greeting *Arms and the Man* with 'admiration and hatred', he was all for eliminating *character* or at any rate banishing it to comedy.[4] The plays of Shakespeare he regarded as tragi-comedy, but, if we agree that Aeschylus and Sophocles, as pure tragedians, do not give us character, what is it they give us in its place? Yeats's answers are inadequate. In 1899 he wrote: 'The theatre began in ritual, and it cannot come to its greatness again without recalling words to their ancient sovereignty.'[5] Certainly the ritual element remains very strong in Greek tragedy, and its words are even stronger—but are they sovereign? I doubt if any scholar or dramatic critic would say so. Why should a play like *Oedipus Rex* carry such enormous weight even in a deplorable translation? In 1916, in praising the Japanese theatre, Yeats wrote: 'It is a child's play become the most noble poetry, and there is no observation of life, because the poet would set before us all those things which we feel and imagine in silence.'[6] To set before us such things is an admirable aim but ought it to be the dramatist's? The theatre, after all, is a noisy place.

Still more give-away is Yeats's comment on his *Deirdre* : 'The absence of character is like the absence of individual expression in wall decoration.'[7] This seems to be the same old modern fallacy which has made painters compare their paintings to carpets and poets attempt a 'pure' poetry where words shall be divorced

[4] Yeats, *The Trembling of the Veil* (1922), Bk 4, in *Autobiographies* (1956), 282–3.

[5] 'The Theatre' (May 1899), sect. i, end of penult. para., in *Ideas of Good and Evil* (1903), *Essays and Introductions*, 170.

[6] *Certain Noble Plays of Japan* (1916), sect. v, in *The Cutting of an Agate* (1919), *Essays and Introductions*, 231.

[7] Letter, 3 April 1902, to Lady Gregory: *The Letters of W. B. Yeats*, ed. Allan Wade (1954), 368. The remark refers to AE's *Deirdre*, not Yeats's.

from their meaning and exalted into mere music—but how inferior to real music!

No one can make a play merely by 'listening to incense', and Yeats, by the grace of God or of Dublin, could not in practice fully implement his themes. All of his *Collected Plays* are worth reading and some remain worth acting. One play which stands on its own, *The Words Upon the Window-Pane*, is indeed highly dramatic; it is significant that this is written in prose and involves two of Yeats's passions—spiritualism and Swift. An earlier play which stands on its own, *Cathleen ni Houlihan*, in which he was assisted by Lady Gregory, must have had a great impact in its day—the day of romantic nationalism. When Yeats wrote later:

> Did that play of mine send out
> Certain men the English shot?[8]

he may or may not have been exaggerating, but at least he was forgetting about wall decoration.

Since Yeats's death Irish intellectuals have gushed about him in print and continued to mock him in private. Which, as he himself might have admitted, is fair enough. For he was both genius and crank, both professional and dilettante. Thus these plays are the work of an amateur playwright—but of a professional poet. He said that he took to drama because he wanted 'clean outline'. Most of us, like George Moore, would find little of clean outline in *The Shadowy Waters*, but it is true that his ventures into drama had a tonic effect upon his lyrics. For this reason any admirer of his lyrics should also read the plays, which contain not only plenty of fine poetry but also, as in *Resurrection* and 'The Cat and the Moon' much of his peculiar philosophy. But it should be remembered that he himself wrote of them:

> Players and painted stage took all my love
> And not those things that they were emblems of.[9]

A sound attitude for a creative writer.

[8] 'The Man and the Echo' (1939), st. 1, written 1938.
[9] 'The Circus Animals' Desertion' (1939), sect. ii, st. 3, written 1937–8.

Dylan Thomas: Memories and Appreciations[1]

Yeats described the poet as one who knows 'that Hamlet and Lear are gay'.[2] No poet of our time was a better example of this than Dylan Thomas. When his first work appeared it was astonishingly new and yet went back to the oldest of our roots—roots which had long been ignored, written off, or simply forgotten. He was not just a poet among poets; he was, as has often been remarked, a bard, with the three great bardic virtues of faith, joy, and craftsmanship—and, one could add, of charity. Many of his poems are concerned with death or the darker forces, yet they all have the joy of life in them. And many of his poems are obscure but it is never the obscurity of carelessness; though I, for one, assumed it might be when I first read his early work in the 1930s. Lastly, all the poems (a rare thing in this age of doubt) are suffused both with a sense of value, a faith in something that is simultaneously physical and spiritual, and with (what is equally rare in an age of carping) a great breath of generosity, goodwill not only towards men but towards all created things.

The next few years will obviously see a spate of writing about Thomas—his vision, imagery, technique, etc.—and the writers will be beset by two distinct and opposite dangers—the danger of trying to equip him too exactly with a literary pedigree and the danger of isolating him as a sport, a Villon figure, a wild man who threw up works of genius without knowing what he was doing. The former mistake has been made for years by various academic critics, often Americans, who have dwelt at length on Thomas's relations to ancient Welsh poetry or to Rimbaud; though he has something in common with both (and though Wales in general and Swansea in particular were the most important factors in his make-up), it should be remembered that he had never read Rimbaud and could not read Welsh. As for the 'wild man' conception, immediately after Thomas's death it was exploited

[1] *Encounter* 2: 1 (Jan. 1954), 12–13; third in a series of memoirs on Dylan Marlais Thomas (1914–53); the other contributors were Daniel Jones, Theodore Roethke, Marjorie Adix, George Barker.

[2] 'Lapis Lazuli', st. 2; frequently cited by MacNeice, as in review 'Great Riches' (Aug. 1950) above.

in its most disgusting and imbecile form by certain of our daily papers. Of course Thomas liked pints of beer (so what? he also liked watching cricket) but he did not write his poems 'with a pint in one hand'; no writer of our time approached his art in a more reverent spirit or gave it more devoted attention. One glance at a Thomas manuscript will show the almost incredible trouble he took over those elaborate arabesques that could yet emerge as fresh as any of the 'woodnotes wild'[3] expected from the born lyric poet. In fact, he *was* a born lyric poet but it was a birthright he worked and worked to secure.

His lyrical gift, though the most important, was only one of several gifts. He had a roaring sense of comedy, as shown in many of his prose works. He had a natural sense of theatre, as was shown not only in his everyday conversation but in those readings of poetry (and his taste, by the way, was catholic) which earned him such applause both here and in the U.S. He was moreover a subtle and versatile actor, as he proved repeatedly in radio performances. *And* he 'took production'. Though his special leaning (as was natural, given his astonishing voice) was to the sonorous and emotional, he enjoyed playing character parts, especially comic or grotesque ones, such as a friendly Raven which he played for me once in a dramatized Norwegian folk-tale.[4] He could even 'throw away' if required to. And in all these sidelines—as in all his verse and prose—there appeared the same characteristic blend of delight in what he was doing and care as to how he did it.

This does not seem to me the moment for analysing Dylan Thomas. He is assured of a place, and a unique one, in the history of English poetry. But, when such a personality dies, his friends are not much in the mood for literary criticism. What we remember is not a literary figure to be classified in the text-book but something quite unclassifiable, a wind that bloweth where it listeth, a wind with a chuckle in its voice and news from the end of the world. It is too easy to call him unconventional—which is either an understatement or a red herring. It is too easy to call him Bohemian—a word which implies affectations which were

[3] John Milton, 'L'Allegro', l. 134, in reference to Shakespeare.

[4] *The Heartless Giant*, broadcast 13 Dec. 1946; cited by Barbara Coulton, *Louis MacNeice in the BBC* (1980), 87–8.

quite alien to Thomas. It is too easy even to call him anarchist—a better word but too self-conscious an attitude. Thomas was an actor—and would that more poets were—but he was not an attitudinizer. He eschewed politics but he had a sense of justice; that he once visited Prague proves nothing as to his leftness or rightness; it merely is one more proof that he thought men everywhere were human. Both in his life and his work he remained honest to the end. This, combined with his talents, made him a genius.

Round and About Milk Wood[1]

Under Milk Wood by Dylan Thomas. (Dent. *7s. 6d.*)
The Doctor and the Devils by Dylan Thomas. (Dent. 10*s. 6d.*)

Dylan Thomas was primarily a lyricist who, to my mind, reached his greatest heights in *Deaths and Entrances*; for sheer lyrical beauty I doubt if any English poem of our time can compare with 'Fern Hill' or 'A Winter's Tale'.[2] He had strict views not about poetry in general but about that poetry which he had been born to write and, while he admired such simple and comparatively pedestrian poets as Edward Thomas and Andrew Young[3] he always insisted that his own poetry must be 'poetic'. Not for him the flat but accurate description, the urban or industrial background, the social or moral comment, the witty understatement, the 'contemporary situation', the didactic or satirical exposition of any 'world view' or philosophy. 'Lions and fires of his flying breath'—those were his sole concern in his poems. And his diction was as strictly controlled as his content; just as he would not and could not mention such a thing as a gasometer, so he eschewed the topical word and the slang word. Once, when we were reading each other our latest poems, he remarked to me

[1] *London Magazine* 1: 3 (April 1954), 74–7; title supplied from typescript at the Humanities Research Center, Austin, Texas.

[2] 'A Winter's Tale' and 'Fern Hill' (both 1945) were both first collected in *Deaths and Entrances* (Feb. 1946).

[3] Edward Thomas (1878–1917) and the Rev. Canon Andrew J. Young (1885–1971) were more prosaic in their nature poetry than Dylan Thomas, lacking his lyric lilt, for which see MacNeice's memoir immediately above.

that for *him* it would be impossible to use the word 'pub' in a poem. Which perhaps is amusing in view of the fact that our popular press has described him as the Poet of the Pubs.

While I deplore the malice that seems to have activated certain journalists in making such an emphasis, it would be a bad mistake to swing to the other extreme and present Dylan Thomas as a type of 'beautiful but ineffectual angel' or even, as some of his compatriots seem to be doing, as primarily a good Christian whose memory would be defiled by any reference to human weaknesses.

Thomas in some respects possibly had not fully grown up but he had a mature common sense; he also, together with his lions and fires, had a fine amount of good rich earth in him. This earthy side showed itself in his conversation and, if much of that conversation did take place in pubs, it should be remembered that a pub is both a more traditional and a more sympathetic institution than the bar of the Ritz—or than a literary party. Thomas possessed what Yeats attributed to Synge, an appreciation of 'all that has edge, all that is salt in the mouth',[4] but he was a vastly better mixer than Synge. It is this side of him which found expression in such prose works as *Portrait of the Artist as a Young Dog* and, more recently, in his radio play *Under Milk Wood*. Compared with his best poems these may be minor works but they are excellent work of their kind and quite unlike anything else I know of by a contemporary. Where else can you find such a natural lack of both bitterness and self-pity, such a natural blend of slapstick and reverence, of fantasy and flesh-and-blood?

Under Milk Wood, recently broadcast for the first time in an exquisite production by Mr Douglas Cleverdon,[5] is a very different affair from the lyrics and, being of a lighter and more popular *genre*, will appeal, I imagine, to a wider audience. Yet it is obviously by the same hand and contains in diffusion much that the lyrics had concentrated. For example a late poem, 'Lament',

[4] Yeats, 'J. M. Synge and the Ireland of his Time' (1910), sect. ix, in *The Cutting of an Agate* (1912), *Essays and Introductions*, 326.

[5] T. Douglas Cleverdon (1903–), publisher and BBC producer, not only produced Thomas's 'play for voices' on the Third Programme 25 Jan. 1954, but also directed its first stage productions in Edinburgh and London 1955 and in New York 1957.

the autobiographical portrait of the same sort of randy, rowdy old man so much fancied by the later Yeats, contains the lines:

> I skipped in a blush as the big girls rolled
> Ninepin down on the donkeys' common

and also:

> And the sizzling beds of the town cried, Quick![6]

This old man obviously lives in the Milk Wood world, a world more conscious of love than of anything else except death. Whether Thomas would ever have made a dramatist (in the ordinary sense) is doubtful, partly because he showed no sign of being able to handle a dramatic plot (*Under Milk Wood* has less in common with a play proper than with a work like *Spoon River Anthology*)[7] and partly just because the main motive for his characters must always have been the lust to possess and be possessed (which with him, as with D. H. Lawrence, is also a spiritual motive). He did however have two great dramatic assets, his dialectical view of life and his interest in the spoken word. This is why *Under Milk Wood* needs to be performed as well as read.

Too many critics (and there will be many more!) have approached Thomas's works either as if they were Holy Writ and could only be ascribed to 'inspiration' or as if they were the hard-born offspring of innumerable esoteric 'influences'. The answer to the 'Holy Writ' critics is that Thomas did *not* write 'as the bird sings'; he was a most painstaking craftsman. The answer to the others is that he was quite unread in Welsh or other foreign literatures but that he spent his childhood in a setting which was both childlike and traditional and had, among other things, a father who spontaneously uttered such phrases as 'happy as the grass is green'.[8] As for his reading, he was saturated in the Bible and in folk tales, revelled in *Mayhew's London*[9] and used to relax with space fiction. In his early poems the constellations of imagery

[6] 'Lament' (1951), from sts 1 and 2.

[7] (1915) by the American poet Edgar Lee Masters (1868–1950)—a miscellany of character sketches in a small town.

[8] 'Fern Hill', st. 1: '. . . happy as the grass was green'.

[9] Henry Mayhew (1812–87), *London Labour and the London Poor* (1851–64), a treasure-house of vivid social history reporting on the low life of London.

are for my taste often too centrifugal but throughout his work there are many single images which at first sight may look very daring but which in fact are the same in kind as the clichés we all use daily—'It's raining cats and dogs', 'Aunt Alice has green fingers'.

Though he read a great deal of English poetry I find him nearer to the folk world than to the bookish world. Think of such carols as 'I saw three ships . . . ' or 'The holly and the ivy', such folk songs as 'Nottamun Town' or, in the nursery rhyme sphere, the hyperboles of the 'Derby Ram' or the strange glamour even of some of the riddles.

> White bird featherless
> Flew from Paradise . . .

says the nursery rhymer, meaning snow; an image which goes back to the ninth century.

Thomas's most recent publication in book form was *The Doctor and the Devils*, a film scenario based on the Burke and Hare murders. This *is* dramatic (though it should be noted that the story line was supplied by Mr Donald Taylor) and would make a most effective film if film-makers were not so afraid of upsetting the public.[10] And I am not, by the way, at all sure that Mr Taylor is right when, assuming that this film, if ever made, will be vastly different from Thomas's screenplay, he also assumes that it will be thereby much improved thanks to 'the varied contributions made by many hands and the constant polishing and alteration that is the *integral* (italics mine) part of the work'. I should have thought, the closer they stuck to Thomas, the better; constant polishing too often means emasculation. Be that as it may, this screenplay makes very good reading; Thomas had a fancy for the macabre, which here he was able to indulge. And there are many lines which illustrate, if not his lyrical, at least

[10] Donald Fraser Taylor (1920?–66), the film producer of Strand Film Co., hired Thomas to write for documentary films 1942–5. *The Doctor and the Devils* (1953), a feature script on the Edinburgh body-snatchers Burke and Hare, commissioned by Taylor, was written by Thomas 1944–7 in collaboration with Taylor, but later sold to the Rank Organization and never produced because of Rank's decision to avoid horror and murder subjects. Callum Mill rewrote it as a stage play and produced it in Glasgow 1961 and at the Edinburgh Festival 1962. See Paul Ferris, *Dylan Thomas* (1977), 181–2, 205, 214–15; Jacob Korg, *Dylan Thomas* (1965), 174. Obituary of Taylor is in *Screen World*, 18 (1967), 240.

his conversational genius: 'When one burns one's boats, what a very nice fire it makes.'

In *Under Milk Wood*, on the other hand, which is a far richer work, the conversational and lyrical geniuses have got together. Much of it is wildly funny, some of it is deeply moving.

Luxuriant scene-painting alternates with a crisp and dovetailed dialogue. There are old gags—the postman's wife steaming open the letters—and ballady songs:

> But the one I loved best awake or asleep
> Was little Willy Wee and he's six feet deep[11]

but they all have the Thomas touch. 'I will lie by your side like the Sunday roast'[12] says the lovesick draper.

> Knock twice, Jack,
> At the door of my grave[13]

says the dead sailors' tart. It is all about love (in Thomas one can't call it sex) and then, once you turn the medal, it is also all about death. Mr Walter Allen was right in calling this a 'daedal dance'.[14] Llarregub *is* a Welsh village—but a Welsh village transmuted by a maker of myths. And painted by the same hand that painted the dead Ann Jones, whose room had contained 'a stuffed fox and a stale fern'. But 'a monstrous image blindly Magnified out of praise' and a number of monstrous, and yet very human, images had built up that poem until

> The stuffed lung of the fox twitch and cry Love
> And the strutting fern lay seeds on the black sill.[15]

[11] Polly Garter in *Under Milk Wood* (1954), 60.

[12] Mr Mog Edwards in ibid., 7.

[13] Rosie Probert in ibid., 77–8.

[14] 'What he [Thomas] was out to express was the daedal dance of life. He did so in *Under Milk Wood* on a larger scale than anywhere else': 'William Slater' (pseud. of Walter Allen, 1911–), 'Look and Listen', *New Statesman and Nation* 47: 1196 (6 Feb. 1954), 159. Mr Allen, a friend of MacNeice, has written to me that at the time he was 'the *New Statesman*'s radio critic. Louis knew this. . . . I was afraid my cover might have been broken. In fact, who William Slater was was widely known at the BBC.'

[15] 'After the funeral (In memory of Ann Jones)' (1938), ll. 16–17, 39–40.

Endless Old Things[1]

The Letters of W. B. Yeats. Edited by Allan Wade. (Hart-Davis. 63*s*.)

A great deal of Yeatsiana has been pouring out since Yeats's death; there is perhaps a moral here for those who wrote him off as finished before World War I. His poet's progress remains a most peculiar one—that of a man who started with every sort of limitation and ended, a lonely public figure, on a pedestal as high as the famous Pillar in O'Connell Street on which Nelson (who was never an Irishman) still stands, with one arm and one eye. To a younger, more socially conscious, generation he was sometimes a suspect figure; George Orwell, like Stephen Spender, found 'sinister implications' in his philosophy and concluded that 'The relationship between Fascism and the literary intelligentsia badly needs investigating, and Yeats might well be the starting-point.'[2] At first sight this looks plausible, but the more we know of Yeats the man, the more we shall probably feel that 'Fascism' here is a red herring. Of course he did use the word himself as when in 1933 he wrote that he was trying to 'work out a social theory which can be used against Communism in Ireland—what looks like emerging is Fascism modified by religion'.[3] And he did back General O'Duffy for a short time, but by 1937, when the General was in Spain with his Blue Shirts, Yeats was writing that the last thing he wanted was to see him 'back in Ireland with enhanced fame helping "the Catholic front"'. And he added: 'I don't know on which side a single friend of mine is, probably none of us are on any side.'[4]

This vacillation between would-be Fascism and neutrality (encouraged by an ignorance of history) may be characteristic of Yeats but is also characteristic of his country. Orwell's essay on

[1] *New Stateman and Nation* 48: 1230 (2 Oct. 1954), 398.

[2] 'George Orwell' (pseud. of Eric Arthur Blair, 1903–50), 'The Development of W. B. Yeats', *Horizon* 7: 37 (Jan. 1943), in *The Collected Essays, Journalism and Letters*, eds Sonia Orwell and Ian Angus, 4 vols (1968), ii. 273, 276.

[3] Letter, April 1933, to Olivia Shakespear: *The Letters*, ed. Wade, 808.

[4] Letter, 11 Feb. 1937, to Ethel Mannin: *Letters*, 881.

him shows little perception either of the simple peculiarities of Ireland as a whole or the more complex peculiarities of that Protestant minority to which Yeats, like many another ardent nationalist, belonged. For Yeats, though he thought Bernard Shaw 'a barbarian of the barricades',[5] would have agreed in his bones with Shaw's statement in his preface to *John Bull's Other Island*:

> When I say that I am an Irishman I mean that I was born in Ireland, and that my native language is the English of Swift and not the unspeakable jargon of the mid-nineteenth century newspapers . . . I am violently and arrogantly Protestant by family tradition; but let no English government therefore count on my allegiance: I am English enough to be an inveterate Republican and Home Ruler.[6]

Yeats too was born and bred Protestant (which in Ireland does imply both violence and arrogance) and, whatever his flirtations with the Cabbala, the Upanishads, and so on, and however great and understandable his envy of Maud Gonne's conversion to Rome, his motto to the end was 'No Surrender'.[7]

All this one could read between the lines of his poems but those lines are so reverberant that it shows more clearly in his letters where, if not unbuttoned (Yeats was never a Bohemian) and if not even unguarded (he was always too much on the attack), he is at least less purged of those trivial concerns and naïveties which give away a man's true position. All of his admirers must be most grateful to Mr Wade for having so carefully collected and edited these give-away letters which evoke a reaction like that of Yeats himself in 1909:

> My dear Wade, I am amazed at your bibliography. . . . The thing is of great value to myself and now you have made it possible I shall re-read endless old things I had thought never to see again.[8]

The bibliography referred to was begun by Mr Wade about 1898; this series of letters runs from 1887 to 1939. Endless old things?

[5] Letter, 7 Oct. 1907, to Florence Farr: *Letters*, 500.

[6] (1904), Preface to 1st edn (1906), in *The Complete Prefaces of Bernard Shaw* (1965), 442: corr. 'the mid-XIX century London newspapers'.

[7] Yeats identified with Protestant Ireland of the Anglo-Irish ascendancy, whose Orange motto was 'No Surrender'—deriving from the siege of Derry 1689. See J. G. Simms, *Jacobite Ireland 1685–91* (1969); Tony Gray, *The Orange Order* (1972), 36–7.

[8] Letter, 15 Jan. 1909, to Allan Wade: *Letters*, 523.

Indeed yes; and how very revealing they are! Some of these have already appeared in print, for instance letters to Katharine Tynan published last year and edited by Roger McHugh; but Mr Wade's 900 pages not only cover the poet's adult life but, owing to the diversity of his correspondents which called forth different parts of him, do compose a solid and credible portrait.

Readers of his later poems would be prepared for a certain astringency, but it is pleasant to be reminded by these letters that the sting was in him so early. Yeats may have invented the Celtic Twilight, but he himself is far from crepuscular when he is concerned with personal gossip or 'theatre business' or money. Not only wit but humour shows itself early and even on subjects which might have been sacrosanct to him:

> A sad accident happened at Madame Blavatsky's lately, I hear. A big materialist sat on the astral double of a poor young Indian. It was sitting on the sofa and he was too material to be able to see it.[9]

In a letter to AE he makes a revealing comparison of their characters:

> The antagonism, which is sometimes between you and me, comes from the fact that though you are strong and capable yourself you gather the weak and not very capable about you, and that I feel they are a danger to all good work. It is, I think, because you desire love. Besides you have the religious genius to which all souls are equal. In all work except that of salvation that spirit is a hindrance.[10]

Yeats himself could not be accused of suffering weaklings gladly and, whether he desired love or not, he certainly was not prepared to let it submerge him. The voice in these letters grows warmer as the years pass but it is worth noting that his men friends are never addressed by their Christian names.

General ideas are few; they changed less than his tastes did. And when he risks an historical generalization it is often a rash one: 'it flashed on me that the Coming of Allegory coincided with the rise of the Middle Class.'[11] But Yeats himself repeatedly denounced 'abstraction' and it is not general ideas, or beliefs, that

[9] Letter, 12 Feb. 1888, to Katharine Tynan: *Letters*, 59.

[10] Letter, 8 Jan. 1906, to George Russell: *Letters*, 466.

[11] Letter, 4 Dec. 1902, to Lady Gregory: *Letters*, 386: corr. 'it flashed upon me . . . '.

make a poet; even in his Introduction to *A Vision* (which at other times he implies is a *Credo*) he describes his circuits of sun and moon, etc., as 'stylistic arrangements of experience comparable to the cubes in the drawings of Wyndham Lewis and the ovoids in the sculpture of Brancusi'.[12] What appears in him to be generalization is often in fact merely stylization. He writes in 1889: 'With me everything is premeditated for a long time'[13] and he went on meditating the same things through the decades till his symbols began turning into persons and the real people he knew into symbols. What he wrote in a late poem about the theatre applies to his art in general, and perhaps to his life:

> Players and painted stage took all my love,
> And not those things that they were emblems of.[14]

Certain other late poems, such as 'The Municipal Gallery Revisited' and 'Beautiful Lofty Things', show him changing an occasional subject-matter into mythology—'All the Olympians; a thing never known again.'[15] To mythologize your friends like this you perhaps have to be somewhat remote from them. If Maud Gonne had married Yeats, what would have happened to his poetry?

These letters demand our attention because Yeats was a 'major' poet. It is a term I dislike but it is useful to indicate the gulf between him and such a poet as AE (just as George Herbert can be called 'major' where Christina Rossetti is 'minor'). The word 'major', however, does carry certain implications of bulk, depth and width, and when we start trying to find these qualities in Yeats, we shall find they have changed in his hands. He *is* wide in a sense—but not the usual sense. He *is* deep in a sense—but not the usual sense. Conclusion: this is a major poet—but a very, very odd one. It will probably be many years yet before critics can get this oddity in focus. In the meantime one can hazard that Yeats is an extreme example of the artistic principle that a man's limitations can be turned into assets. If he had not been both tone-deaf and comparatively ignorant of prosody, he would not have evolved

[12] Introduction to *A Vision* (1937; 1956), 25: corr. 'drawing . . . and to the ovoids'.

[13] Letter, 31 Jan. 1889, to Katharine Tynan: *Letters*, 107.

[14] 'The Circus Animals' Desertion' (Jan. 1939), ii, st. 3; also quoted at the end of the review 'Yeats's Plays' (Nov. 1952) above.

[15] 'Beautiful Lofty Things' (1938), last line.

his extraordinary verbal music. If he had not had extremely bad eyesight, he might have got stuck, as Tennyson sometimes did, in merely brilliant description. If he had had a better education, it should have saved him from the follies of astrology—yet those follies are responsible for some of his finest work. His drawbacks became steps forward—but this miracle would not have happened if he had not been uncommonly hard-working and, according to his own strange lights, honest.

I Remember Dylan Thomas[1]

A great deal has been written, much of it hastily, about Dylan Thomas and his work since his death last November.[2] Inevitably a rather strange picture has emerged, or rather several strange pictures according to the particular emphasis made by the various authors. For instance, some of the memorial articles emphasized the 'Bohemian' qualities of the man, while others emphasized the dedicated, not to say religious, character of the poetry.

To relate the poetry to the man or even to relate his poetry to his other works, such as the prose stories and the radio play *Under Milk Wood*, is certainly by no means easy. What does stick out a mile at first sight is that both as a man and as a writer he was an 'original'—and a very fascinating one. As I was not one of his oldest friends (nor did I ever see him in his family circle) I am much less qualified than several others to assess Dylan Thomas as a whole; all I propose to do here is to record certain personal impressions and reminiscences of him.

[1] *Ingot* (Steel Co. of Wales), Dec. 1954, 28–30, accompanied by three portraits and two poems ('And death shall have no dominion' and 'In my craft or sullen art') and by this editorial epigraph: 'It is just a year since he died—the ebullient young Welshman whose poems were written, he said, "for the love of Man and in praise of God". In this contribution a fellow poet, himself distinguished in the world of letters, gives his personal recollections of Dylan Thomas as he knew him.'

[2] 9 Nov. 1953 in New York City, from alcoholism and from morphine administered by Dr Milton D. Feltenstein: see Paul Ferris, *Dylan Thomas* (1977), chap. 14, 299–309.

I met him first in 1941,[3] so I never knew the slim and faunlike figure which was painted by Augustus John and which, when he was barely grown up, had swept through literary London rather like a forest fire. But, comparing notes with those who knew him from his Swansea schooldays, I realize that Dylan was always the same Dylan, perhaps because in a sense (a good sense) one could say that he had never entirely grown up. He retained the alertness and bubble of a child and also a child's integrity (which last does not of course imply that children are, on the business plane, 'reliable').

My first meeting with Dylan was not a great success, partly because from the start he was somewhat suspicious of me (as a member of a group of poets whom certain critics, tendentious or naïve, had placed in an opposite camp to himself) and partly because I unwittingly confirmed this suspicion by telling him that in his last book there were five—or maybe it was six—poems which I thought were really good. Dylan obviously considered that such a specification of the good poems was patronizing. Owing no doubt to this suspicious attitude on his part (which I expect I reciprocated) I cannot remember that at this first meeting he impressed me with that exuberant yet subtle charm which subsequently was to draw me like a magnet whenever I heard that he was in London. Nor can I remember that on that occasion he did or said anything funny.

The Cause of Comedy

Later I realized that, quite apart from his astonishing lyrical gifts, the man was a comic genius—and the cause of comedy in others. Hugh Griffith has compared him to a royal jester who was also the royal bard.[4]

Dylan in his poetry avoided all mention of the urbanized, industrialized, commercialized, politically conscious and embittered modern world. In his life he spent a great deal of time

[3] Error for 1938 (18 Oct.), in a BBC broadcast 'The Modern Muse' with Auden, Spender, Day-Lewis, and MacNeice: Constantine Fitzgibbon, *The Life of Dylan Thomas* (1965), 255.

[4] Hugh Griffith (1912–80), the popular Anglo-Welsh actor, in a memorial BBC programme *Dylan Thomas: a Portrait of the Man as his Friends Remember Him*, broadcast 11 Dec. 1953.

in the very hubs of modernity—London, New York, Chicago, Rome, or Prague—but, however mass-produced the setting and however sophisticated, desiccated, or professionally spiteful the company, he brought his own atmosphere with him. At chromium-plated bar or literary party the weather would change in a wink—the slick generalizations, the cynical poses, would be punctured, and the people would remember they were human.

I gradually got to know Dylan through my work for the BBC, where as a producer I realized that he was a god-send to radio. His famous 'organ voice' was already well known in straightforward readings of verse, but the same voice, combined with his sense of character, could be used for all sorts of strange purposes. I cast him (and was never disappointed) in a variety of dramatic parts, including that of a funereal but benevolent raven in a dramatized fairy story.[5] What was more surprising was that he was capable, when necessary, of *underplaying*—but the need for this had to be made clear to him. He took his radio acting very seriously and between rehearsals would always keep asking if he were giving satisfaction. He was a joy to have around in the studio, causing a certain amount of anxiety to the studio managers, who could never be sure that he would speak into the right microphone, and a great deal of delight to the rest of the cast who particularly admired the queer little dance steps that he always performed (it seems quite unconsciously) while broadcasting.

A Cricket Match

It was not, however, till he had worked for me several times in radio shows that Dylan and I spent a whole day together in recreation. We met about eleven in the morning and went immediately to Lord's cricket ground, having typically failed to check that the match which we wished to see was still on. The match being already over, we decided to switch to the Oval but, since Dylan remembered that somewhere between Lord's and the Oval there was a very good underground wine bar, we drank champagne (which neither of us could afford) till three, after which

[5] *The Heartless Giant*, broadcast 13 Dec. 1946: cited by Barbara Coulton, *Louis MacNeice in the BBC* (1980), 87–8; see 'Dylan Thomas: Memories and Appreciations' above with n. 4.

we did go to the Oval, where he poured out much curious lore about cricketers' private lives, all of it funny and most of it, I fancy, true.

We ended the day eating Portuguese oysters. After that we were always, I think, at ease with each other.

'A small round comfortable man' was his own description of himself; sometimes he said 'a little fat fool'—but the fool should have a capital F. Like many great clowns he was under the average height but he held himself very straight and his presence was not only dignified but, less expectedly, dapper. He dressed in an 'artistic' way but never looked slovenly, just as he spoke and acted unconventionally and yet somehow gave an impression not only of ingrained courtesy but of a certain respect for conventions; only the conventions were his own. His Byronic curls were usually countered by a neat bow tie; the Byronic open neck would have shocked him.

When he entered any company there was always a re-grouping and Dylan became the centre. Once in the centre, he began to twinkle and sparkle, talking through a cigarette that grew from the corner of his mouth until, since he had no time to knock the ash off, it became one pure stream of ash sustained in the air by a miracle. His eyes (gooseberry and bogwater) would roll with a warm-hearted mischief, his bulbous body and face would seem light and gay as a balloon at a children's party, and his words would caper. He warmed any room that he entered; which is why since his death certain places seem unnaturally chilly to his friends.

A special event, the memory of which I cherish largely owing to Dylan, was a very formal poetry reading given for the present Queen Mother (she was then the Queen) in the Wigmore Hall, London, shortly after the war.[6] The readers were an assortment of well-known actors and poets, including Edith Evans, Edith Sitwell, John Gielgud, John Masefield, and Walter de la Mare. We were instructed to report behind the stage a quarter of an hour before the Queen was due to arrive. Everyone turned up on time except Dylan, who was never noted for punctuality.

[6] 'A Recital of Poetry of the Past and Present, the Poet Laureate Presiding', Wigmore Hall, London W1, on Tuesday, 14 May 1946, at 3 p.m.—printed programme at the Humanities Research Center, Austin, Texas.

We waited for him in something of an anxious gloom which was enhanced by the sombre dark suits all the men had put on for the occasion. At last he appeared, with about two minutes to spare, and walked straight up to a full-length mirror which stood in the middle of the green room.

What he saw in it appeared to shake him. He was wearing, if I remember correctly, a thick shaggy sports jacket of striking blue tweed, a couple of half-buttoned cardigans and beneath them a green flannel sports shirt, a tie pulled askew, and, to cap—or rather to underpin—it all, a pair of blue and white check trousers such as comedians used to wear in music halls.

Having postured and puzzled for some moments in front of the mirror, he turned and looked us up and down in all our sabbatical formality and then came to me with a proposition. As only one reader was to be on the stage at a time, would I lend him my suit whenever he had to go on? Looking at Dylan and comparing his shape and structure with my own, and then looking at our distinguished colleagues before whom we should have to go through these lightning changes I suggested (though I felt rather a cad) that he would look much better the way he was. So he stayed the way he was and he did look much better—and much better, I might add, than most of us. The first poem he had to read was Blake's 'Tiger, Tiger', which had been preceded by a much longer poem, 'Kubla Khan', read by an actor who was a specialist in verse reading. This actor stood in the wings as Dylan started, and after hearing the first two words his eyes began to pop and he began to look at his wrist-watch. Blake's 'Tiger' came over in all his solemn magnificence . . . but I have never heard a poem read so slowly.

When Dylan came off, the actor went up to him and said: 'Dylan! Do you know you took longer over "Tiger, Tiger" than I did over "Kubla Khan"?' Dylan, appearing slightly nettled (but perhaps he was only mock-nettled) replied: 'Well, I took it as fast as I could.' He had in fact, as so often, brought a welcome spice of fantasy into the proceedings, blended (his own special blend) with an equally welcome spice of earthiness.

Joy of Life

A born actor and mixer, he exuded an immense joy of life and also a love of human beings. In the last few months before his death a shadow appeared to creep in, partly I think owing to ill health and partly to the imminence of that spiritual crisis which most poets have to face when they are approaching forty; I first noticed this in him in New York in April of last year. But the few times I met him in London during the following summer he was as wonderful company as ever. We spent one afternoon together at Lord's, where we were lucky enough to see Denis Compton make a century[7]—and that in his earlier, more carefree and impudent manner. Dylan was delighted since Compton, whose batting he described as 'feline', was one of his favourite cricketers. I think he felt a natural affinity with anyone in any sphere who combined, as Dylan himself did, originality and technical skill, who was so well versed in his craft that he could stand the rules on their head.

The last time I saw him was the night before he flew to America, a country which amused him vastly. But for once he seemed sad to go. He kept harping on a headline in one of the day's papers (a peculiarly comic and wicked double meaning); he mimicked a simple millionaire whom he'd met that morning; he tried to persuade an old Cockney to sing us a ribald ballad. Then he felt ill and went home, carrying himself straight as usual, with the usual warm smile over his shoulder.

Sometimes the Poet Spoke in Prose[1]

Quite Early One Morning. By Dylan Thomas. 240 pp. (New York. New Directions. $3.50.)

When a writer dies, it is much better that he should be rhapsodized over in the tradition of the wake than damned with faint praise in the tradition of the smart obituary. Since, however, the near-apotheosis of Dylan Thomas (1914–53) has already provoked

[7] Denis C. S. Compton (1918–), record-breaking batsman for England and Middlesex, as well as an Association footballer, retired 1957 from both careers because of a knee injury: *The Oxford Companion to Sports & Games*, ed. John Arlott (1975), 182.

[1] *New York Times Book Review*, 19 Dec. 1954, 1; editor's subtitle: 'From Squib to Satire—a Collection of the Varied Writings of Dylan Thomas'.

certain spiteful reactions in England, it is time we got certain things straight. Thomas had extraordinary gifts, but these, like most gifts, were limited. Not everything he wrote was a masterpiece. He was first and foremost a lyrical poet, secondly a writer of comic or nostalgic prose, thirdly (a dubious third) a dramatist, fourthly (a poor fourth) a critic.

For anyone who knew him, such works as *Portrait of the Artist as a Young Dog* and the magnificent pageant (which is not really a play) of *Under Milk Wood* perpetuate a most important and lovable side of him, which is hardly to be glimpsed in the poems. It is this side that is to the fore in this group of radio talks, sketches and squibs, and appreciations of literature, these last being chiefly of interest as proving his catholic tastes, his innocence and his integrity.

Just as Thomas was not afraid in his poetry to be fastidiously, even archaically, 'poetic', so in his prose he was never afraid to be hail-fellow. Too many writers nowadays are terrified of appearing either sentimental or vulgar. Thomas could use the word 'lovely', he could also dredge up a joke from the world of the seaside picture postcard. J. M. Synge thought that the finest works of art are produced when the ordinary man is 'rising into dream' or the dreamer coming down out of it.[2] This makes a good description of the best pieces in this book such as 'Holiday Memory' and 'A Story', in both of which, as so often in Thomas, it is the memory of childhood that promotes just this wedding of dream and reality.

Take the picture of the uncle in 'A Story'. 'As he ate, the house grew smaller; he billowed out over the furniture, the loud check meadow of his waistcoat littered, as though after a picnic, with cigarette ends, peelings, cabbage stalks, birds' bones, gravy; and the forest fire of his hair crackled among the hooked hams from the ceiling.'[3] And the title piece, which breaks into verse, anticipates *Under Milk Wood*:

[2] '. . . what is highest in poetry is always reached where the dreamer is leaning out of reality or where the man of real life is lifted out of it. and [*sic*] in all the poets the greatest have both these elements that is they are supremely engrossed with life and yet with the wildness of their fancy they are always passing out of what is simple and plain.' From a Synge notebook, quoted in Introduction to J. M. Synge, *Collected Works*, 4 vols. (1962–6), i: *Poems*, ed. Robin Skelton (1962), xv.

[3] *Quite Early*, ed. cit. (American expanded edition), 45.

Clara Tawe Jenkins, 'Madame' they call me,
  an old contralto with her dressing gown on.
And I sit at the window and I sing to the sea,
  for the sea doesn't notice that my voice has gone . . .

  I am Mrs Ogmore-Pritchard and I want another snooze.
Dust the china, feed the canary, sweep the drawingroom floor.
  And before you let the sun in, mind he wipes his shoes.[4]

This is a child's world. It is also a wit's world. This wit of Thomas is of first importance (I have heard him improvise a string of brilliant parodies of a dozen contemporary poets), but the wit is balanced by humour. Thus he greatly admired Joyce Cary's *The Horse's Mouth*, a book condemned to me once with typical gustophobia by a leading British editor as merely 'facetious and sentimental'. But he would have none of P. G. Wodehouse and explains that 'a truly comic, invented world must live *at the same time* as the world *we* live in'.[5] As for the human warmth, even in satirical pieces like his shattering 'A Visit to America', it keeps bubbling up through the satire. For Thomas was no smiler with a knife, he did not belong in the clinics of Aldous Huxley, he belonged in the Mermaid Tavern. 'And, twittering all over, old before their time, with eyes like rissoles in the sand, they are helped up the gangway of the home-bound liner by kind bosom friends (of all kinds and bosoms).'[6] There is here no unkindness, no prejudice; European lecturers, American lecturees, they all get it full in the face—but *it* is a custard pie.

Mr Robert Graves in a recent broadcast denied that poetry is or should be entertainment.[7] Thomas thought differently: 'I think there's an inverted snobbery in being proud of the fact that one's poems sell very badly. *Of course*, nearly *every* poet wants his poems to be read by as many people as possible. Craftsmen don't put their products in the attic.'[8] Hence he did not pigeonhole his

[4] Ibid., 18–19. Cf. *Under Milk Wood* (London, 1954), 16, and ed. Douglas Cleverdon (1972), 26: 'And before you let the sun in, mind it wipes its shoes.'

[5] 'A Dearth of Comic Writers', *Quite Early*, ed. cit., 158; Thomas's italics.

[6] 'A Visit to America' (1953), ibid., 233.

[7] 'Poems should not be written, like novels, to entertain or instruct the public; or the less poems they.'—Robert Graves, 'A Poet and his Public', Home Service broadcast, 20 Oct. 1954, in the *Listener* 52: 1339 (28 Oct. 1954), 711.

[8] 'On Poetry' (1946), *Quite Early*, ed. cit., 193; Thomas's italics.

works, as Graves does, into public or potboiling (prose) and private (poetry). Wales and childhood (the two seem sometimes interchangeable) appear in his prose with a different, often a comic, emphasis, but it is the same Wales and the same child re-creating it.

The same is true of his voice when he gave public readings; it remained the same (and what an astonishing voice) whether he was reading a cryptic poem on childbirth or a stretch of O'Casey-like prose or a bit of Welsh character dialogue. Mr Hugh Griffith,[9] a well-known Welsh actor, recognizing in Thomas a peer, has compared him to the Fool in Shakespeare; the Fool is a professional entertainer, but he has also his private depths from which his wits catch fire.

So let us get it straight and forget all the wishwash about 'neo-romanticism' (Thomas would not be pleased to hear he was still being bracketed with Barker and Treece)[10] and remember that here was a man who was both a craftsman and a mixer. He was liable to get drunk on words (some of the early poems suffer from this), but in the best pieces in this book that sheer horse sense, which inspired the Fool in *King Lear* and which enabled Thomas to chat with any chance comer, binds his fantasies to earth, while that earth itself, since he fulfils the requirements of Synge, is lit with the lights of dream. Stage lighting? Certainly. That makes it no less real.

Thomas would never have subscribed to the surrealist doctrine that the poet is 'a modest registering machine'.[11] He knew that the poet is a maker rather than a piece of litmus paper; he knew that poetry is organic and not mechanical. And as for modesty, while painfully aware of the poet's and his own limitations, he was proud to admit that poetry 'is the most rewarding work in the world. A good poem is a contribution to reality.'[12]

[9] For Hugh Griffith's remark, see article 'I Remember Dylan Thomas' (Dec. 1954) above with n. 4.

[10] George Barker (1913–   ) and Henry Treece (1911–66) were 'Neo-Romantics' who used hyperbole and other bombastic forms of expression in their verses, like the poets of 'The New Apocalypse' in the 1940s.

[11] See MacNeice's 'Subject in Modern Poetry' (Dec. 1936) above with n. 48.

[12] 'On Poetry', *Quite Early*, ed. cit., 192.

Poetry and the Age / Inspiration and Poetry[1]

Poetry and the Age by Randall Jarrell. (Faber and Faber. 18*s.*)
Inspiration and Poetry by C. M. Bowra. (Macmillan. 21*s.*)

'But taking the chance of making a complete fool of himself—and, sometimes, doing so—is the first demand that is made upon any real critic; he *must* stick his neck out just as the artist does, if he is to be of any real use to art.'[2] This quotation, from a chapter 'The Age of Criticism' in Mr Jarrell's new book, indicates that its author is far from being a typical American critic. And in this book he certainly does stick his neck out (and would that Sir Maurice Bowra did the same!) and sometimes, many English readers may think, does make a fool of himself, as in what most of us will consider his overpraise of William Carlos Williams and in many other passages where he not only loads his contemporaries with superlatives but mentions them in the same breath as Dante, Shakespeare *et al.* All of which I find admirable, especially as Mr Jarrell puts his cards on the table; he will tell you exactly which poems by anyone he thinks are his best, he will quote from these abundantly, *and* he will tell you exactly why he admires what he quotes. There are few contemporary critics, either here or in America, who show such enthusiasm, honesty or wit.

In a world where most of the sceptics are cold pike and most of the enthusiasts melting jellyfish, Mr Jarrell stands out as someone well equipped not only with a heart but with several grains of salt. For example, while persuasively eulogizing Wallace Stevens (a poet for whom I have always had a blind spot) he makes his eulogy more acceptable to the unconverted by admitting Stevens's characteristic weaknesses—weaknesses which have grown on him in his last book, *Auroras of Autumn*, where, as Mr Jarrell says, 'the green spectacles show us a world of green spectacles'.[3] With another poet whom he greatly admires, Marianne Moore, he is again aware of the weaknesses ('as if one could represent a stream by reproducing only the stepping-stones one crossed it on')[4] and allows himself, as often, one or two

[1] *London Magazine* 2 :9 (Sept. 1955), 71–4.
[2] Jarrell, 85; Jarrell's italics.
[3] Ibid., 134; *The Auroras of Autumn* (1950).
[4] Jarrell, 177.

wisecracks, saying of her earliest poems: 'Butter not only wouldn't melt in this mouth, it wouldn't go in.'[5] And in an excellent article on Robert Frost he rightly maintains that 'Frost's best-known poems with a few exceptions, are not his best poems at all' and that 'so far from being obvious, optimistic, orthodox, many of these poems are extraordinarily subtle and strange, poems which express an attitude that, at its most extreme, makes pessimism seem a hopeful evasion'.[6]

The chapter from which I took my opening quotation is a magnificent example of well canalized bad temper. Mr Jarrell himself apologizes for it but anyone conversant with that extraordinary American phenomenon, the university course in 'Creative Writing' or with the jargon-constipated pseudo-scientific utterances of so many American (and some British) literary critics should feel that such an apology is not needed. As Mr Jarrell says, 'a great deal of this criticism might just as well have been written by a syndicate of encyclopedias for an audience of International Business Machines'.[7] His polemic is extremely funny—and also, I think, unanswerable. As he says, 'more and more people think of the critic as an indispensable middle man between writer and reader', whereas, as he also says, 'the work of art is as done as it will ever get, and all the critics in the world can't make its crust a bit browner'.[8] And we surely must all agree with his parting piece of advice: 'remember that works of art are never data, raw material, the crude facts that you critics explain or explain away.'[9]

One of Mr Jarrell's best chapters is on Whitman, a poet who, as he perceives, is too often admired or disapproved of for the wrong reasons. Whitman of course is a freak but American poetry (or, for that matter, poetry) cannot be considered apart from him and, since he *is* a poet, one must look in him for the poetic differentia and not assume, in Mr Jarrell's words, that he is 'a sort of Balzac of poetry, whose every part is crude but whose whole is somehow great'.[10] For in no poetry can the parts be divorced from the whole. Many of us can remember, when we were undergraduates, laughing our heads off at Whitman's absurder passages but, as Mr Jarrell says, 'surely a man with the most extraordinary feel for language, or none whatsoever, could

[5] Ibid. [6] Ibid., 37. [7] Ibid., 73.
[8] Ibid., 90. [9] Ibid., 92. [10] Ibid., 107.

have cooked up Whitman's worst messes'.[11] It is this 'extraordinary feel for language' which Mr Jarrell illustrates by his usual method of quotation (he quotes as aptly as Matthew Arnold) and thereby proves his main contention 'that Whitman was no sweeping rhetorician, but a poet of the greatest and oddest delicacy and originality, and sensitivity, so far as words are concerned'.[12] 'So Far as Words are Concerned' might almost be the title of this book; every page brings out the forgotten truism that words are what poems are made of.

It is interesting to compare Mr Jarrell with Sir Maurice Bowra. He too is an enthusiast, and one with enormous knowledge and a most unusually wide taste, but somehow more often than not he fails to convey his enthusiasm. It is not only that he is often writing about authors, e.g. Russian or Portuguese, whom most of us cannot read in the original; even on Milton or Thomas Hardy I find that he fails to express what he must indubitably feel about them. This is largely a question of style—and of personality. We all know that Sir Maurice is very much a person but, unlike Mr Jarrell, his writing lacks personality and does not convince because it does not provoke. Neither of them uses jargon but, when reading sentences of Sir Maurice's such as 'They convey in flawless form the all-absorbing excitement which comes from untrammelled inspiration',[13] I feel like throwing him away and taking up instead such a Marxist jargoneer as Christopher Caudwell.[14] Caudwell makes heavy reading but at least he is trying to get beneath the surface.

Two of the more interesting chapters in Sir Maurice's book deal, respectively, with Horace and Walter Pater, two writers who to-day are out of fashion or, as Mr Jarrell's enemies the portentous critics would put it, no longer Important. Sir Maurice shows sound sense in recognizing (what many fail to) that Pater, 'a typically English empiricist', had his share of sound sense too.[15] Pater was not a great writer but he was very perceptive and had the courage of his perceptions; as Sir Maurice says, 'In an age when Ruskin, despite prodigious knowledge and an extraordinary sensibility,

[11] Ibid., 110. [12] Ibid., 107–8.

[13] The songs of Gil Vincente, the Portuguese poet (*c.*1470–*c.*1540): Bowra, 111.

[14] 'Caudwell' (pseud. of Christopher St John Sprigg, 1907–37), remembered for two works of Marxist criticism, *Illusion and Reality* (1937) and *Studies in a Dying Culture* (1938).

[15] Bowra, 203.

spoiled much of his work by views which soon became antiquated, Pater saw further and kept closer to permanent elements in imaginative experience'.[16] As for Horace (who *was* a great writer) it is refreshing to find a critic surmounting those Romantic prejudices against him which most critics have by now surmounted in regard to the English Augustans (not but what Horace is far more romantic than Pope). But I wish Sir Maurice had tried to analyse more deeply Horace's 'originality'. He *is* original—but not according to most twentieth century definitions.

The most disappointing chapter in Sir Maurice's book is his first one, on 'Inspiration and Poetry'. Back in the Nineteen-Thirties 'inspiration' was a word which many of us wanted to scrap. But, luckily for all of us, the thing behind the word is something completely unscrappable. Sir Maurice is right when he says, 'Paradoxically this condition is one of joy'[17] but I wish he had dug deeper; as it is, he has called in evidence various dissimilar poets but has failed, I feel, to lay hands on or at any rate to transmit the evasive common denominator. It is interesting to know that Pushkin 'insisted that the impulse to write is properly called "enthusiasm" and the critical activity "inspiration"',[18] and to be reminded that Keats began with Shelley's untenable ultra-Romantic theory and dropped it before he wrote the Odes; but (and perhaps again this is merely because the critic shows his hand) I find more illumination in Mr Jarrell's unashamedly rhetorical statement: 'A good poet is someone who manages, in a life-time of standing out in thunderstorms, to be struck by lightning five or six times; a dozen or two dozen times and he is great.'[19]

Lost Generations?[1]

Poetry Now, Edited by G. S. Fraser. (Faber. 15*s*.)
Mavericks. Edited by Howard Sergeant and Dannie Abse. (Editions Poetry and Poverty. 6*s*.)

This is not a proper book-review (how can a few hundred words cover four score contemporaries?) but rather a protest,

[16] Ibid., 207. [17] Ibid., 7. [18] Ibid., 12.
[19] Jarrell, 136, at end of chapter 'Reflections on Wallace Stevens'.

[1] *London Magazine* 4: 4 (April 1957), 52–5.

a sticking-out-of-the-neck. There have been too many anthologies of contemporary verse and much in their introductions has been either dull or ridiculous. There are exceptions, of course. *New Lines*[2] last year was worth while because it confined itself to nine poets (rightly or wrongly presented as a 'group') who deserved to be better known. *Poetry Now*, edited by G. S. Fraser, who ought to know better, includes seventy-four poets, not one of them represented by more than two poems—oh, the inevitable bittiness of it! Mr Fraser's Preface, except when he is quite rightly praising his compatriots, e.g. Robert Garioch and W. S. Graham,[3] is chiefly of interest for the holes that can be picked in his somewhat half-hearted generalizations. He says nothing richly absurd, such as Yeats and the Apocalyptics used to offer us,[4] so let us take our lesser toothpicks to him. He finds, for instance, in the early 1950s 'the emergence of a really new attitude to poetry';[5] this I shall return to but, taking him chronologically and on ground I know better, I must first challenge what he says about the 'Thirties' while thanking him for not treating the 'Thirties' as a dirty or non-U phrase.

This game of pigeonholing literary generations has gone too far. Mr Fraser classifies the 30s poets, mentioning particularly Auden and myself, as 'Augustans', who were succeeded by 'romantics' (who, unlike us, built poems from 'images rather than statements') and also by some university poets, such as Mr Heath-Stubbs, who 'represented a kind of Back-to-Literature movement'.[6] This last would imply that Mr Auden, whose danger has always been bookishness, was somehow opposed to literature or less well up in it than Mr Heath-Stubbs. But more irritating is the Augustan–Romantic antithesis. Posterity may find our generations closer to each other than we care to think. Such a line as 'Beginning

[2] *New Lines*, ed. Robert Conquest (1956), featured nine poets of 'The Movement': Elizabeth Jennings (1926–), John Holloway (1920–), Philip Larkin (1922–85), Thom Gunn (1929–), Kingsley Amis (1922–), D[ennis] J[oseph] Enright (1920–), Donald Davie (1922–), Robert Conquest (1917–), John Wain (1925–).

[3] 'Robert Garioch' [Sutherland] (1909–81) and W[illiam] S[ydney] Graham (1918–86), Scottish poets.

[4] See 'Apocalyptics' in 'An Alphabet of Literary Prejudices' (March 1948) above, para. 1.

[5] G[eorge] S[utherland] Fraser, 'Preface', *Poetry Now* (1956), 22.

[6] Ibid., 17–18.

with doom in the bulb, the spring unravels'[7] could well have been written by Auden or Day-Lewis. And

> Some life, yet unspent, might explode
> Out of the old lie burning on the ground[8]

could have been written by Spender. Take another 'Romantic':

> The lion-tamer Discipline was mauled,
> Sentiment bitten by its own charmed snake,
> The lariat Intellect lassoed itself . . . [9]

Surely these lines are more on the 'Augustan' ticket? And, on the other hand, what is so Augustan about

> Or look in the looking glass in the end room—
> You will find it full of eyes,
> The ancient smiles of men cut out with scissors and kept in mirrors?[10]

But it is unfair to go by odd lines alone; let us look at whole books. Mr Fraser, though more cautious, writes as if he would support the extraordinary statement by Mr Heath-Stubbs, introducing another anthology, that 'the poets of the 1930s . . . tended to limit themselves to the apprehension of social and political realities'.[11] It partly depends on how one defines 'social' (language itself is 'social' after all), but anyhow is it not obvious that the early Auden was at least as influenced by Freud (let alone Groddeck!) as by Marx? It is just these psychologists that real social-politicos have always blamed for their concentration on the individual and on personal relationships. But, such 'influences' aside, look at the books of the 30s. Spender's *Poems* (published 1933) shows at least fifteen out of thirty-three poems that could not be called 'social' or 'political' at all. In Auden's *Look Stranger* (1936) the figure is at least fifteen out of thirty-one; in my *The*

[7] Dylan Thomas, 'I, in my intricate image' (1935), i, st. 2: *The Poems*, ed. Daniel Jones (1971), 108.

[8] Dylan Thomas, 'I have longed to move away' (1935), st. 2: op. cit., 44.

[9] A mixed imitation, combining 'Romantic' and 'Augustan' elements, of Auden's birthday poem for Christopher Isherwood, 'August for the people' (1935), st. 10; untraced quotation: perhaps a concoction by MacNeice to challenge G. S. Fraser.

[10] This quotation, a characteristic leg-pull, is by MacNeice himself—'Perseus' (Aug. 1934), st. 2: *Collected Poems*, 25.

[11] John Heath-Stubbs, 'Introduction', *The Faber Book of Twentieth Century Verse*, eds Heath-Stubbs and David Wright (1953), 30.

Earth Compels (1938) at least thirteen out of twenty-four. And in all these books most of the poems I am conceding to the social-and-or-political category remain, to my mind, highly personal (sometimes too much so) and often even 'romantic'—in the tradition of that earlier 'social-political' poet, Shelley. As for Day-Lewis's *From Feathers to Iron* (1931) this whole sequence of twenty-nine poems is concerned with his wife having a baby; true, there are images in it drawn from the social or political—or industrial—spheres, but these are strictly subsidiary. One more word on the dirty 'Thirties'. Mr Fraser implies that with that generation their 'attitude to life has, probably, never fundamentally changed since they were undergraduates'.[12] Let him come up and see us some time.[13]

Now to skip the 'New Romantics' and Co. and come to the 'Movement' led by Messrs Amis and Wain, that 'really new attitude to poetry' which, in Mr Fraser's words, represents an 'ousting of the bohemians by the pedants'.[14] Mr Fraser includes a number of these poets among his seventy-four elect and from some of them such as Miss Jennings, Mr Holloway and Mr Larkin,[15] I would have welcomed far more than the meagre ration here allowed them. Yet these better ones strike me as *individual* poets; it is not their group characteristics which make them interesting. The chief characteristics of the 'Movement', we are often told, are neatness and lucidity. But are these so new in our time? To mention just one name from many, what about the late Norman Cameron?[16] Surely any professional poet, grouped or ungrouped, dated or dateless, ought to be able, *when he chooses*, to be neat or lucid, or both; just as any professional draughtsman should be able to command, when he needs them, the old-fashioned tricks of perspective. But what distinguishes the 'Movement' as a group from a poet like Robert Graves[17] is that the latter's neatness is always a means to an end. As individuals

[12] Fraser, op. cit., 26–7.

[13] MacNeice's note: 'I am not for a moment, of course, conceding that "social and political realities" are in any way alien to poetry. See Ancient Greece *passim*.'

[14] Fraser, op. cit., 22.

[15] The first three poets of *New Lines*.

[16] [John] Norman Cameron (1905–53), a poet of exact form, lucidity, and wit.

[17] Robert [von Ranke] Graves (1895–1985), also a poet of exacting form and wit, though markedly mythopoeic.

then, we must welcome some of these New Liners, but as a group or a Movement, well, let them go. And behold, they go—with what docile arrogance, with what lowered but polished sights; roped together, alert for falling slates, they scale their suburban peaks—the Ascent of C3.[18] And the banner with the strange device will soon droop on the cairn of red bricks. 'Snaffle and Curb'—but see Roy Campbell about that.[19]

Yes, it would be a bleak prospect if everyone wrote Movementese. As Mr Howard Sergeant says in *Mavericks*, their 'concentration upon form is, in fact, concentration upon the same two or three forms and rhythms, repeated to the point of tedium';[20] and, as Mr Danny Abse says, 'With the Movement poets the reader hardly ever receives the impression that the poem has seized the poet.'[21] Let us be thankful that there are still poets among us who get such creative seizures (oh how very non-M of them!), poets like W. S. Graham, Laurie Lee, Norman MacCaig and W. R. Rodgers;[22] they are non-M in several ways, being subtle in form, rich in content, sometimes simultaneously sensuous and mystical, yet with wits or wit enough to control both their sense impressions and their spiritual urges. Messrs Sergeant and Abse have assembled nine other non-M poets (four of them represented in *Poetry Now*) whom they have chosen as being writers who 'are unafraid of sensibility and sentiment, who are neither arid nor lush'.[23] None of these nine, as yet, seems to me comparable with the four mentioned above but I find in Mr Jon Silkin[24] that same sympathetic blend of physical and mystical, while I am glad that Mr John Smith in 'Walking on Stilts' is not

[18] Lower goal or 'Parnassus': ironic echo of Auden and Isherwood's play *The Ascent of F6* (1936, rev. 1937).

[19] Roy Campbell (1901–57), 'On Some South African Novelists' in *Adamastor* (1930), *The Collected Poems*, 3 vols. (1949–60), i. 198:

> You praise the firm restraint with which they write—
> I'm with you there, of course:
> They use the snaffle and the curb all right,
> But where's the bloody horse?

[20] Sergeant (1914–), 'Introduction II', *Mavericks* (1956), 13.

[21] Abse (1923–), 'Introduction I', 9.

[22] Four regional poets of various rich idioms taken directly from nature, more seemingly spontaneous than the 'New Liners'.

[23] Abse, op. cit., 10.

[24] Jon Silkin (1930–) was represented by five poems in *Mavericks*, 57–64.

afraid to be lilting nor Mr Michael Hamburger in 'Islands' afraid of the musical cliché. All the same I must confess that I find more that I like in *New Lines*. Mr Robert Conquest in his Introduction to that anthology claimed that it illustrated 'the principle that poetry is written by and for the whole man'.[25] An admirable principle—only his writers as a group do not illustrate it; some of them do, as individuals. So two final morals. (1) No one poet can ever appear at his best in an anthology. (2) No poet, unless negatively, will benefit for long from subscribing to any group theory of poetry.

Louis MacNeice Writes . . . [on *Visitations*][i]

I feel delighted and honoured that the Poetry Book Society have chosen my book, *Visitations*. This is the first book of short poems I have published since 1948. In between I have published *Ten Burnt Offerings* (ten long poems which were experiments in dialectical structure) and one very long poem, *Autumn Sequel*, the point of which was missed by most of the book-reviewers; it was 'occasional' but not casual, being an attempt to marry myth to 'actuality'. While writing these longer pieces I was incapable of writing short ones. When the lyrical impulse did return, this interval of abstention, it seems to me, had caused certain changes in my lyric-writing—I naturally hope for the better. It is hard to put labels on one's own work but I like to think that my latest short poems are on the whole more concentrated and better organized than my earlier ones, relying more on syntax and bony feature than on bloom or frill or the floating image. I should also like to think that sometimes they achieve a blend of 'classical' and 'romantic', marrying the element of wit to the sensuous-mystical element.

So much for my own poems. Now, as to 'poetry in general'—there is, strictly speaking, no such thing, every poem being a concrete individual. Still, one *can* generalize about these individuals.

[25] Conquest, 'Introduction', *New Lines*, xiv.

[i] *Poetry Book Society Bulletin*, no. 14 (May 1957), [1]. *Visitations* (Faber, 10 May 1957) was the Summer Choice 1957 of the Poetry Book Society.

I hold that poetry, far from being a release of gas, is more like a precision instrument—one that can be used where that other precision instrument, science, is completely and for ever useless. That is, I agree with the late Christopher Caudwell that poetry is inevitably subjective—but this need not imply either imprecision or isolation; as Caudwell pointed out, the poet retires into his inner world thereby to re-establish communion with his fellows (not all his fellows, of course, but a worthwhile number).[2] So long as hunger remains as real as bread, this will be an important activity. As for my own preferences, I like a poem to exist on more than one plane; I hope this is true even of my own 'poems of place' (there are several such in *Visitations*) which, superficially, are merely descriptive pieces. I also think a poet should not be afraid of being thought either sentimental or vulgar. Lastly, we should always remember that, while a good poem cannot but be an artefact, it is also certainly an organism.

James Joyce's World / Letters of James Joyce[1]

James Joyce's World by Patricia Hutchins. (Methuen. 30*s*.)
Letters of James Joyce. Edited by Stuart Gilbert. (Faber. 42*s*.)

James Stephens, quoted by Miss Hutchins, said of *Finnegans Wake*: 'This book is not written in prose, it is written in speech . . . speech moves at the speed of light, prose moves at the speed of the alphabet.'[2] So to catch their Joyce as it flies is a hard task for his interpreters. Shaw, Yeats, and Joyce, all three Irish and, in their quite different ways, only too adept at word-play, were three gigantic cranks. But they all were quick on the laugh and it seems an irony that their works should now be at the mercy of the evergrowing tribes of humourless scholars, of Quintilian

[2] 'In poetry itself this [struggle with Nature] takes the form of man entering into emotional communion with his fellow men by retiring into himself': 'Caudwell' (pseud. of Christopher St John Sprigg), *Illusion and Reality: A Study in the Sources of Poetry* (1937, new edn 1946), 125.

[1] *London Magazine* 4: 11 (Nov. 1957), 73–5.

[2] In a broadcast of readings from *Finnegans Wake* (1939), Jan. 1947, cited by Hutchins, 221–2.

G. Grinder on his campus or little Plucky Jim making red bricks without straw in the hope they will add up to a doctorate. At this rate June 16 will soon be not Bloomsday but Boresday.[3] Joyce wrote in a letter to his grandson (see Mr Gilbert's collection): 'The devil mostly speaks a language of his own called Bellsybabble which he makes up himself as he goes along but when he is very angry he can speak quite bad French very well though some who have heard him say that he has a strong Dublin accent.'[4] Neither Jim nor Grinder can really understand this accent—like many a stranger to Dublin they cannot tell the wind in the reeds from the tongue in the cheek—and so they dehydrate Joyce, destroying both the devil and the child in him.

I do not include either Mr Gilbert (who knew Joyce) or Miss Hutchins (who didn't) in this condemnation, but both their new books are disappointing. Miss Hutchins's is less interesting than her earlier *James Joyce's Dublin*, which it inevitably overlaps (where else was his world anyway?), and even the photographs in it are duller than those in the earlier book, such as Noel Moffett's intriguing and apposite sand patterns; while Mr Gilbert, who in 1930 published, with Joyce's approval, a fascinating study of *Ulysses*, appears, in editing these letters which are most of them themselves pretty dull, a halfhearted and inattentive editor, labouring less from love than because he is stuck in a groove. Both these books make sticky reading. Joyce's letters, like Yeats's, are largely concerned with royalties, book reviews, etc.—many of them are written to his patroness Harriet Shaw Weaver[5]—and are often formal and stiff; as with Yeats, decades of acquaintance could not bring him down to Christian names. The redeeming moments are those of arrogance, as when at the age of nineteen he writes to Ibsen, or of playfulness as in several letters to Frank

[3] Bloomsday, 16 June 1904, is the day on which the action of *Ulysses* (1922) is set; its anniversary has been widely celebrated by Joyceans.

[4] Postscript to letter, 10 Aug. 1936, to Stephen Joyce: *Letters*, ed. Gilbert (1957, corr. 1966), 387. Vols ii and iii of Joyce's *Letters* were edited by Richard Ellmann (1966).

[5] Editor of the *Egoist* 1914–19 who published Joyce's *A Portrait of the Artist as a Young Man* as a serial and as a book (1916, 1917), as well as the early episodes of *Ulysses* : she helped issue *Ulysses* as a book, was a source of funds over many years, and became Joyce's literary executrix; yet Miss Weaver did not appreciate *Finnegans Wake*. See Jane Lidderdale and Mary Nicholson, *Dear Miss Weaver: Harriet Shaw Weaver 1876–1961* (1970), *passim*.

Budgen. As for Miss Hutchins, too much of her book is irrelevant conjecture, as when she writes about Joyce's stay in Bognor where he was toiling at *Work in Progress* (the italics that follow are mine): '*If* he asked his daughter . . . the name of the big red house on the corner, and she told him it was a House of Rest for Women, he *probably* gave his short laugh and asked her if there wasn't one next door for weary writers.' Nor is it illuminating to have every building described, in Zurich or Paris, that Joyce, a chronic nomad, ever stayed in for a week or two. Miss Hutchins probably enjoyed this paper chase but Solomon had the right word for it.

Still let us be grateful for the crumbs. It is interesting to learn from Miss Hutchins that the youthful Joyce used to play charades, dressing up for instance as Carmen, and also liked a game which was a play on place names, e.g. 'Harold's Cross because Terenure' (Terry knew her); and that at the same period he always asked in the Irish National Library for *The Illustrated London News*. The books an author read are usually more revealing than the houses he lived in and I am glad that Miss Hutchins, in a footnote, details the periodicals Joyce bought during two months in 1929: '*The Baker and Confectioner*, *Boy's Cinema*, *The Furniture Record*, *Poppy's Paper*, *The Schoolgirl's Own*, *Woman*, *Woman's Friend*, *Justice of the Peace*, *The Hairdressers' Weekly*'. Still better are some of the rare declarations she quotes from the master's lips, e.g. 'I hate generalities' (take note, all jargoneers and boreocrats) or the magnificent 'If *Ulysses* isn't fit to read, life isn't worth living.'[6] But it must be remembered that Joyce, like T. S. Eliot, is a paradox; both have drawn on colloquial speech—and Joyce at least is a master of written dialogue—but both remain essentially literary and neither of them could be described as a sparkling conversationalist. Which is why Dr Richard Best, onetime National Librarian of Ireland, said to me of Joyce—and he said the same of Synge: 'I didn't think he was important; you see, he had no conversation.' It is an irony that the greatest celebrant of Dublin should have been so lacking in the Dubliner's most famous virtue or vice.

Mr Gilbert's collection of letters also contains its memorable sentences. Joyce writes in 1902, still under age, to Lady Gregory: 'I have found no man yet with a faith like mine.'[7] To Grant

[6] Hutchins, 149 and 139.

[7] Letter, Nov. 1902: *Letters*, ed. Gilbert, 53.

Richards about *Dubliners* in 1906: 'I believe that in composing my chapter of *moral* (italics mine) history in exactly the way I have composed it I have taken the first step towards the spiritual liberation of my country.'[8] In 1917 to Ezra Pound: 'Unfortunately, I have very little imagination.'[9] In 1919 to Harriet Shaw Weaver on *Ulysses* : 'I confess that it is an extremely tiresome book but it is the only book which I am able to write at present.'[10] And to Frank Budgen on the chapter of parodies in *Ulysses* : 'As I told you a catchword is enough to set me off.'[11] And to Miss Weaver again in 1921: 'I have not read a work of literature for several years. My head is full of pebbles and rubbish and broken matches and lots of glass picked up "most everywhere".'[12] And there is a very interesting explanation, to Budgen, of the last episode in *Ulysses* and its four cardinal points: 'It begins and ends with the female word *yes*.'[13] And, beginning his yet stranger experiments, he parodies 'St Patrick's Breastplate': 'Complications to right of me, complications to left of me, complex on the page before me, perplex in the pen beside me, duplex in the meandering eyes of me, stuplex in the face that reads me.'[14] There is also a commentary—'just forty-seven times as long as the text'—on a very short passage of *Finnegans Wake*.[15] These letters bring out his dedication and his innocence, his formality and his inconsequence (he admits that his sympathy with Vico was heightened because he too, unlike most Italians, feared thunderstorms). When he writes to his family there is also a tenderness. Above all, though he was never 'engagé' and appears to have loathed Irish politics, like all politics, there is the regional fixation; at the age of fifty-five he writes from Paris: 'But every day in every way I am walking along the streets of Dublin.'[16] Or, as he put it in one of his last puns, 'the Finn again wakes'.[17]

[8] Letter, 20 May 1906: op. cit., 62–3.

[9] Letter, 9 April 1917: ibid., 101.

[10] Letter, 20 July 1919: ibid., 128.

[11] Letter, Michaelmas 1920: ibid., 147; on episode 15, 'Circe' (or 'nighttown'), *not* episode 14, 'Oxen of the Sun' (parodies).

[12] Letter, 24 June 1921: *Letters*, 167.

[13] Letter, 16 Aug. 1921: ibid., 170.

[14] Letter, 16 Nov. 1924, to Harriet Shaw Weaver: ibid., 222.

[15] Commentary 'Twilight of Blindness Madness Descends on Swift'—Letter, 23 Oct. 1928, to Harriet Shaw Weaver: ibid., 273–4.

[16] Letter, 6 Aug. 1937, to Constantine P. Curran: ibid., 395.

[17] Letter, 8 Feb. 1940, to Frank Budgen: ibid., 408.

Joyce's own work will be waking when his scholiasts have all gone to sleep—if they are not asleep already over their typewriters.

The Variorum Edition of the Poems of W. B. Yeats[1]

The Variorum Edition of the Poems of W. B. Yeats. Edited by Peter Allt and Russell K. Alspach. (Macmillan. £4.4*s.*)

Yeats, that indefatigable painter of lilies and chewer of cuds, obviously required a variorum edition and anyone who likes to study poets at their 'stitching and unstitching'[2] must be grateful to Dr Alspach and the late Peter Allt for supplying this. But I doubt the claim made by Professor T. R. Henn in his tribute to Allt that 'it may well be that we have here the supreme mine from which we can now draw more understanding of the name and nature of great poetry; perhaps even of the nature of that most elusive thing, the coming and the quality of inspiration'.[3] We should be far more likely to find such a mine in a study of the *manuscript* versions of these poems, but this, as the editors point out, 'must await their coming to light through the years'. As it is, this is a labour of love which involved the collating of 101 books and also of many periodicals. Inevitably a great many of Yeats's emendations, thus lavishly and meticulously spread before us, are merely to such things as spelling and punctuation (in neither of which he felt at home) but there is still a great deal which is more deeply significant and which, if it does not throw light either on 'the name and nature of poetry' or on 'inspiration', does throw light upon Yeats himself ('It is myself that I remake' he wrote)[4] and on his development both in ideas and technique.

[1] *London Magazine* 5: 12 (Dec. 1958), 69, 71, 73, 75. See MacNeice's 'Yeats at Work' (March 1963) below for review of a study of Yeats's manuscripts or work-sheets.

[2] Yeats, 'Adam's Curse' (1902), st. 1.

[3] T. R. Henn, 'George Daniel Peter Allt', *Variorum Edition* (1957), xiii.

[4]

> The friends that have it I do wrong
> When ever I remake a song,
> Should know what issue is at stake:
> It is myself that I remake.

— [untitled] piece (1908) not in the definitive edition: *Variorum Edition*, 778.

Nearly twenty years ago when I was writing a study of Yeats I was much perplexed by his habit of revision. Thus taking 'The Sorrow of Love' (1892) and comparing the original with the version published three decades afterwards, I conceded that this later version was truer to the later Yeats, who in his search for 'more of manful energy' had turned his back upon languor ('those outlines of lyric poetry that are blurred with desire and vague regret') but I also suggested that 'perhaps this poem *ought* to be languid'.[5] It is not a question of whether the later Yeats had the right thus to supersede the early Yeats but rather a question of 'by their fruits . . . '. Does the later version succeed in its kind as well as the original succeeded in its kind? Or should not Yeats perhaps by 1925 have left the poem of his youth to sink or swim and written a brand-new poem on what he might think was the same subject? Still, this variorum edition has made me more sympathetic to many of the changes. 'One is always cutting out the dead wood,' he wrote in 1927[6] and when this is his main object and not also the grafting of an Irish yew on a willow, when it is a question of detail, of rhythm or epithet (epithets by the way tend to be cut by half), his changes are usually for the better. For example, 'Before earth took him to her stony care' is obviously a better line than 'Before earth made of him her sleepy care'.[7] And the poem about the frustrated lover which once ended:

> Although the rushes and the fowl of the air
> Cry of his love with their pitiful cries

ends better with the subtraction of the pathetic fallacy and the addition of rhetoric:

> O beast of the wilderness, bird of the air,
> Must I endure your amorous cries?[8]

A striking example of a poem of the 'nineties which he gradually improved is 'Red Hanrahan's Song about Ireland'. This consists of three stanzas and the first line of each passed through three stages, thus:

[5] MacNeice, *The Poetry of W. B. Yeats* (1941, 1967), 71, without italics.

[6] 'Preface' (Jan. 1927) to *Poems* (1895), new rev. edn (1927): *Variorum Edition*, 848.

[7] 'The Man who Dreamed of Faeryland' (1891), st. 1: *Variorum Edition*, 126.

[8] 'He thinks of his Past Greatness . . . ' (1898), ll. 11–12: ibid., 177.

1 (a) Veering, fleeting, fickle, the winds of Knocknarea . . .
1 (b) O tufted reeds, bend low and low in pools on the Green Land . . .
1 (c) The old brown thorn trees break in two high over Cumann's Strand . . .
2 (a) Weak and worn and weary the waves of Cummen Strand . . .
2 (b) O tattered clouds of the world, call from the high Cairn of Maive . . .
2 (c) The wind has bundled up the clouds high over Knocknarea . . .
3 (a) Dark and dull and earthy the stream of Drumhair [Dromahair?] . . .
3 (b) O heavy swollen waters, brim the Fall of the Oak trees . . .
3 (c) The yellow pool has overflowed high up on Clooth-na-Bare . . . [9]

Both the first version (all triplets) and the second version (all vocatives) are obviously more sentimental than the third which is in every way stronger and (or because) more concrete. Thorn trees breaking in two hit the reader harder than reeds bending low—let alone 'low and low'. The waves of Cummen Strand, unless it is untypical of the strands of Connaught, must have been off colour that day of the first version. The verb 'bundled', unlike the verbs and adjectives in the earlier versions, is not only right but arresting. 'The yellow pool' we can accept as a particular and therefore a real pool, unlike all those pluralities of winds, reeds, clouds, etc. Lastly, the rhythm in the third version is much better than in the other two.

So far I have been considering the earlier poems. Among these is the dedicatory poem about 'the bell-branch full of ease', first published in 1890 and still being revised in the 1920s.[10] Here the changes are largely dictated by a change in Yeats's attitude to Irish nationalism. To take only one example (I treated this poem at some length in my book on Yeats),

> I tore it from the green boughs of old Eri,
> The willow of the many-sorrowed world . . .

[9] (1894), ll. 1, 6, 11: ibid., 206–7.

[10] 'The Dedication to a Book of Stories selected from the Irish Novelists' (1891), st. 3: ibid., 130. Cf. MacNeice, *The Poetry of Yeats*, 72–3.

becomes

> I tore it from the barren boughs of Eire,
> That country where a man can be so crossed . . .

While I prefer the later more bitter version (Yeats said of it, 'Even in its rewritten form it is a sheaf of wild oats'),[11] this may well be merely wisdom after the event. After all, the Celtic Twilight may have been bad history but at least it *made* history. Still, it is natural for a poet—there have been some notable examples in our time—to try to exorcize his own early commitments. When we come to Yeats's middle period, we find less drastic surgery. For example, that fine austere poem 'No Second Troy' (1910) was never changed apart from the insertion of one comma. 'Pardon, old fathers . . . ' on the other hand (1914) had to be altered because Yeats had made a mistake as to which side the old fathers stood on at the Boyne. One short poem of this period, 'The Scholars', was, I think, spoilt in its revision some fourteen years later. Where the second verse had begun

> They'll cough in the ink to the world's end;
> Wear out the carpet with their shoes
> Earning respect; have no strange friend;
> If they have sinned nobody knows . . .

Yeats rewrote it:

> All shuffle there; all cough in ink;
> All wear the carpet with their shoes;
> All think what other people think;
> All know the man their neighbour knows.

Why?

Of poems later still, 'Leda and the Swan' (1924) was much improved by revision, especially in its opening which, though never bad, was transformed into a triumph, but in this case the revision followed quickly and the poem remains of the same kind.[12] There is nothing so odd about a poet revising his work while his attitude and manner have not altered. It is the earlier cases I have discussed that raise at least two tricky problems. When a poet has developed a new technique, presumably appropriate

[11] Yeats's note after title: *Variorum Edition*, 129.
[12] (1924, 1925, 1928): ibid., 441.

to the subjects he now prefers, how far can he validly impose this on those earlier subjects of his which are now not completely his? Yeats's own answer, in a preface of 1899, is surprising: 'Other revisions are necessary, and he hopes to make them when he is *further from the mood in which the poems were written. . . .*' (italics mine).[13] This raises the second question, that of 'personality' in poetry, about which so much nonsense has been written (on both sides) in our time. A pseudo-impersonality is one of our contemporary bugbears; by substituting other pronouns or nouns for 'I' and 'me' a poet can not only pretend to be far more objective than he is but can guard himself in advance against charges of sentimentality. Such a procedure is alien to Yeats who always had the courage of his ego. All the same he admits that, as he grew older, he became more ready to accept intellectual generalizations. Of course these very soon became part of his ego (readers of *The Vision*[14] might be tempted to add 'And how!'). Still, as with technique, can he validly impose that new part of himself on that anti-intellectual of the 'nineties? He claimed that as he grew older his poetry grew younger, but how can the aged eagle stretch his wings for himself when young?[15] I know no general answer to these questions. My only conclusion about Yeats's revisions (see above) is that sometimes he gets away with them; sometimes not.

A Modern Odyssey[1]

Poems. By George Seferis. Translated by Rex Warner. (Bodley Head. 15*s*.)

English readers have already had several chances to meet the greatest living Greek poet:

[13] 'Preface' (1899) to *Poems* (1895, rev. 1899): *Variorum Edition*, 846.

[14] More accurately, *A Vision* (1925; rev. 1937; 1956).

[15] Echo of '(Why should the agèd eagle stretch its wings?)': T. S. Eliot, *Ash-Wednesday* (1930), i, st. 1.

[1] *New Statesman* 60: 1553 (17 Dec. 1960), 978–9. The greatest then-living Greek poet—writing under the pseud. 'Seferis'—George Sepheriades (1900–71) was Ambassador in London 1957–62; he won the Nobel Prize for literature 1963. He and MacNeice were good friends. For the poet and novelist Warner (1905–86), see MacNeice's 'Poetry To-day' (Sept. 1935) above, n. 98.

I woke with this marble head in my hands . . .
I awoke with this marble head in my hands . . .
I awoke with this marble head between my hands . . . [2]

Mr Rex Warner is a very good translator but sometimes, as in this example, he has been forced into awkwardnesses through trying not to repeat the earlier translators of Seferis—two gifted crews, Messrs Durrell, Spencer and Valaoritis and Messrs Keeley and Sherrard.[3] Still, all these three versions seem to me preferable to the first translation of this line—by Robert Levesque: 'Je me suis réveillé, cette tête de marbre entre les mains . . . '.[4] The demotic Greek of the original has itself a marble quality which English perhaps can get nearer to than French; 'I woke' in the Greek, by the way, takes up just one word. Similarly, two lines down in the same poem, the translations run:

It was falling into the dream as I was rising from the dream . . .
It was falling into the dream as I was coming out of the dream . . .
It was falling into the dream as I rose from the dream . . .

as against: 'Elle plongeait dans le rêve tandis que j'en sortais . . . '

As Seferis's hypnotic effects are largely procured by the repetition of words or phrases (in this line ' . . . óneiro . . . óneiro'),[5] M. Levesque here again loses badly. Yet to all these translators we should be grateful and perhaps most to Mr Warner for the reason that, as its blurb says, 'this volume is the most extensive collection of George Seferis's work to have appeared in English'.

What then is this poet who, writing in a language which, as Mr Warner says, presents even more difficulties than most to the translator, yet keeps inspiring translations and always seems to survive in them? Our first clue is that he was born in 1900 in Smyrna, which is why, as Keeley and Sherrard put it, his poems

[2] A line of *Mythistorema* (1935), sect. iii, in the three successive translations noted by MacNeice in his following sentence: see n. 3 below for the first two translations.

[3] *The King of Asine and Other Poems*, trans. Bernard Spencer, Nanos Valaoritis, Lawrence Durrell, intro. Rex Warner (1948); selected poems in English trans. in *Six Poets of Modern Greece*, trans. Edmund Keeley and Phillip Sherrard (1960), 107–45. (The latter have since published Seferis's *Collected Poems* in three editions [1967, 1969, 1981]).

[4] *Choix de Poèmes*, trans. Robert Levesque (1945), a bilingual edn.

[5] 'Dream'.

taken together 'constitute one long work, a modern Odyssey.'[6] It would be interesting for someone, who has vastly more knowledge than I have of both the countries and both the languages, to make an analytical comparison of Seferis and his Italian contemporary Salvatore Quasimodo, the title of one of whose poems, 'A me pellegrino',[7] might serve for so much of Seferis's work. In both of them the sense of absence seems to become something positive. But Seferis retains what Mr Warner, drawing on a remark made by the poet himself, calls a 'kind of strong poverty',[8] something akin to the Aegean landscape where the bones always show and to the hard Greek light that accuses while it invigorates. I remember Henry Moore arriving for the first time at Piraeus and saying something like: 'This is a sculptor's country but it must be hell for painters.'[9]

Seferis as a young poet was inspired and influenced by Eliot's 'Marina' but in his own work the sea is both more personal and more Greek—which in his own work comes to the same thing. All through his poems flow the images of the voyage—'Their souls became one with the oars and the rowlocks.'[10] And the voyagers keep asking (unanswerable?) questions:

> But what are they looking for, our souls that travel
> On decks of ships outworn . . . ?

The poem which begins like this ends:

> We knew it that the islands were beautiful
> Somewhere round about here where we are groping,
> Maybe a little lower or a little higher,
> No distance away at all.[11]

Which perhaps *is* an answer; on a plane just a shade above or below our own or just round the corner which after all is our

[6] Keeley and Sherrard, 'Introduction', *Six Poets*, 18.

[7] 'To pilgrim Me'. See the bilingual edn, *The Selected Writings of Salvatore Quasimodo*, ed. and trans. Allen Mandelbaum (1960), 158–9. Quasimodo (1901–68) had won the Nobel Prize for literature 1959.

[8] Warner, 'Foreword', *Poems* by George Seferis, trans. Warner (1960), 6.

[9] In early 1951 when MacNeice was still working and living in Athens, Moore arrived for an exhibition of his work at the Zappeion Gallery sponsored by the British Council: see Edward H. Teague, *Henry Moore: Bibliography and Reproductions Index* (1981), 10.

[10] *Mythistorema*, iv, in Warner trans., 14.

[11] Ibid., vii, in op. cit., 19.

own corner, so near and yet so far in fact, lies something which might make sense of both our past and future and so redeem our present.

Past and present are always interfused in Seferis; and his past, of course, is the most notable in Europe. Hence the recurring imagery of broken marbles. Just as in Dylan Thomas the tides in the human body mingle with the tides of the macrocosm, so in Seferis the men and the statues keep turning the tables on each other. It is typical that in one of his most famous poems, *The King of Asine*, he should have chosen as his protagonist—or prototype—not one of the big names but someone

> Unknown, forgotten by everyone, even by Homer—
> Just one word in the Iliad and that word doubtful . . . [12]

'The houses that I had they took from me' a late poem begins.[13] Yet Seferis can create a house from a rotting ship or a ruin and in every case can people it. With the people for whom he has compassion. The heroes. The lost. The anonymous.

Louis MacNeice Writes . . . [on *Solstices*][1]

The poems in *Solstices* were mainly written in 1959 and 1960; in particular, in the spring and early summer of 1960 I underwent one of those rare bursts of creativity when the poet is first astonished and then rather alarmed by the way the mill goes on grinding. Now that I look at the whole collection in cold blood I find that, while it has much in common with my last volume, *Visitations* (1957), fewer of these later poems strike me as forced (in revising I eliminated one or two compulsive bits of trickery) and more of them seem to be 'given'. And the chronic problem of *order* did not seem so difficult as usual. These forty-odd poems include personal lyrics (felt and caught in a flow), personal reminiscences of the war years, a little direct or indirect satire,

[12] *The King of Asine*, in ibid., 71.
[13] 'The House near the Sea', in ibid., 93.

[1] *Poetry Book Society Bulletin*, no. 28 (Feb. 1961), [2], followed therein by MacNeice's poem 'The Truisms' (*Collected Poems*, 507). *Solstices* (Faber, 10 March 1961) was a Spring 1961 Recommendation of the Poetry Book Society.

a few 'travel poems', several sequences and a large number of overt or covert parables; yet, while some deliberately lilt and some deliberately drag, I find that they seem mostly to be scored for the same set of instruments.

Poets are always being required—by the critics and by themselves—to 'develop'. Most critics, however, to perceive such development, need something deeper than a well and wider than a church-door. In certain poets of our time the changes are conspicuous enough; in others, such as Robert Graves, a careless reader might complain that the menu is never altered. To assess one's own development is difficult. I would say of myself that I have become progressively more humble in face of my material and therefore less ready to slap poster paint all over it. I have also perhaps, though I venture this tentatively, found it easier than I did to write poems of acceptance (even of joy) though this does not—perish the thought— preclude the throwing of mud or of knives when these seem called for. Several poems in *Solstices*, e.g. 'Country Weekend', were deliberate exercises in simplicity or at least in a penny-plain technique where fancy rhythms and rhymes would not obtrude too much.

Then of course there is the question of 'commitment'. Some people complained that my long rambling *Autumn Sequel* (1954) was much less committed than its long rambling predecessor *Autumn Journal* (1939); their reasoning seemed to be that the proportion of myth to topicality was much higher in the later work. I do not follow this reasoning. In *Solstices* there is a sequence of four poems suggested by the literature of the Dark Ages; these seem to me just as 'topical' as the poem called 'Jungle Clearance Ceylon' or two that are about the last war. Similarly, when my central image is a windscreen wiper, I feel myself just as mythopoeic as if I were writing about the Grael (though I notice, to my own surprise, that *Solstices* contains practically no allusions to either Graeco-Roman or Christian legend). My own position has been aptly expressed by the dying Mrs Gradgrind in Dickens's *Hard Times*: 'I think there's a pain somewhere in the room, but I couldn't positively say that I have got it.'[2] So, whether these recent poems should be labelled 'personal' or 'impersonal', I feel that somewhere in the room there is a pain—and also, I trust, an alleviation.

[2] *Hard Times, for These Times* (1854), II. ix, para. 40.

That Chair of Poetry[1]

'Even among the Shilluk the structural roles are opposed.' The young anthropologist in a pub on the Woodstock Road was commenting on the candidature of Dr Enid Starkie for the Oxford Chair of Poetry.[2] What he meant was the old cliché: King-makers never become kings. The other anthropologists, all tousled, owl-spectacled and supporters of Dr Leavis[3] ('That man brings tears to one's eyes'), waved their pints and indulged in the sort of quick gossip to be expected from characters in Evelyn Waugh. Since they themselves are more like characters in Kingsley Amis, I marvelled once again at what Oxford does to her captors or infiltrators. Whether they come from the Redbrick enclaves or from Cambridge, they seem to pick up overnight the soft-spoken malice, the ostentatiously throw-away display of inside information, the heavy-lidded thin-lipped irony, the addiction to verbal arabesques, the exquisite verdigris of cynicism, that have traditionally characterized this city of sneering spires.

The anthropologists, who, while deploring my ignorance of African tribal rituals, were prepared to rush into the prosody of Sir Thomas Wyatt, were now playing a saloon-bar game at the expense of, or in homage to (you can never tell which in Oxford), all four of the distinguished candidates: 'A vote for Leavis is a vote for Lawrence . . . A vote for Starkie is a vote for Rimbaud . . . A vote for Gardner is a vote for Chatterley[4] . . . A vote for Graves is a vote for Graves.'[5] At this point the King-maker

[1] *New Statesman* 61: 1561 (10 Feb. 1961), 210, 212.

[2] Enid Mary Starkie (1900?–1970), the distinguished French specialist at Oxford, was 'a force in the politics of Senior Common Room, Faculty Board, and University': obituary in *The Times*, 23 April 1970, 14 f.

[3] F[rank] R[aymond] Leavis (1895–1978), the English critic and educationalist at Cambridge, founder-editor of *Scrutiny*, author of *The Great Tradition* (1948), *D. H. Lawrence* (1955), and numerous other books on English literature and university education.

[4] Dame Helen [L.] Gardner (1908–86), the English Renaissance critic at Oxford, author of *The Art of T. S. Eliot* (1949), editor of John Donne and of Oxford Books, had recently defended Lawrence's *Lady Chatterley's Lover* (1928, 1959) at the 'Chatterley trial' 1959–60.

[5] Robert [von R.] Graves (1895–1985), the British poet and novelist of formal wit and mythopoeia, translator, author of over 137 books including *The White Goddess* (1948), was elected Professor of Poetry at Oxford 1961–6.

herself arrived and asked for an Irish whiskey. She looked fighting fit in bright red corduroy slacks, a bright red beret and a bright blue duffle coat.

Enid Starkie's king-making began ten years ago in the Lewis v. Lewis election.[6] Before that Professors of Poetry got elected without anyone noticing. It was Dr Starkie who introduced the two revolutionary principles that the Professor ought to be (a) a poet and (b) an outsider; this was a shot in the arm for the electorate, but may prove a petard for herself. The Oxford MAs, who have the vote, suddenly found it fun to use it; after all no other Oxford Chair is filled in the same way, with the possible exception of the Lady Margaret Chair of Divinity—but to have a vote for that you need to know something about divinity.

To vote for the Chair of Poetry, an institution described by the outgoing Auden as 'comically absurd',[7] you need not even know the English language. Nor need you to occupy the Chair; nothing in the statutes would prevent the election of a Chinese botanist who would give three lectures a year, in Chinese, on Leninist botany. But Enid Starkie, when first she started her canvassing at the Oxford and Cambridge cricket match (which she attends annually to buy wine for Somerville where she is Reader in French Literature), started a lot of March hares which are still rampaging not only through Fleet Street but Manhattan; and it now seems to matter bitterly who succeeds to the throne once warmed or chilled by John (*Christian Year*) Keble, Matthew (*Dover Beach*) Arnold, F. T. (*Golden Treasury*) Palgrave, A. C. (not F. H.) Bradley, C. D. (not C. S.) Lewis, and W. H. (*Onelie Begetter*)[8] Auden. It all began with la Starkie clutching her brandy in front of the Tavern at Lord's with her back to the cricket, her stockings out-blueing the sky and her blue hat loftily pinnacled at a height of around five feet.

This election has become four-cornered, but it is really One Poet v. the Laity. Other poets, including Betjeman, had been

[6] C[live] S[taples] Lewis (1898–1963), the Cambridge professor of medieval and Renaissance English, v. C[ecil] Day-Lewis (1904–72), the poet-friend of Auden, who held the Chair of Poetry at Oxford 1951–56.

[7] W. H. Auden (1907–73), Professor of Poetry at Oxford 1956–61, 'The Poet as Professor', *Observer*, no. 8849 (5 Feb. 1961), 21: 'Oxford should feel very proud of herself for having anything so comically absurd as a Chair of Poetry.'

[8] 'W.H.', the mysterious dedicatee of Shakespeare's *Sonnets* (1609).

approached; Robert Graves, the only one to accept, is backed, according to the Oxford University Gazette, by 121 Masters of Arts (one more signature than the total secured by Enid Starkie for Auden). As the poll is unlikely to exceed 500, this gives Graves a good start, though the other factions point out that very few of the Gravesmen have anything to do with the Faculty of English. As a slippered champion of Helen Gardner (the slippers were monogrammed) put it: 'The fact that some bureaucratic German scientists were made to sign something after dinner is not going to persuade anyone.'

The 11th-hour incursion of Dr Leavis has confused everyone and led to at least one charge of apostasy. The same slippered Gardnerite alleges that Leavis is supported only by '*Marxisants* Anglo-Catholics'. One of their leaders, 'a philologist who knows about commas in Anglo-Saxon', can be run to earth, it seems, in the Church of St Mary Magdalene 'with Marx sticking out of one pocket, and out of the other the Gospel according to Dr Pusey'.[9] In this atmosphere it is not surprising that Lord David Cecil[10] should have been caused 'great turmoil' and loss of sleep by having to choose between friends, or that Sir Isaiah Berlin,[11] who appears in the Graves list, should say on the telephone that he has 'absolutely no views on the subject whatsoever'.

Some people think that too many Heads of Houses are supporting Graves for his good; it was proved they said, when Macmillan was elected Chancellor, that his opponent suffered from being backed by the academic Establishment. 'From a quarter to nine this morning', said one un-Establishment don, 'I have had to put up with what other people feel.' Such a healthily egocentric resentment might well affect the floating vote; some people feel for instance that Graves, though still the favourite, is being oversold. Also held against him of course are his notorious Clark Lectures at Cambridge.[12] What is held against Leavis is not so

[9] E. B. Pusey (1800–82), the Anglo-Catholic leader of the Tractarian Movement and Regius Professor of Hebrew at Oxford.

[10] Lord [E. C.] David [G.] Cecil (1902–86), the author, writing on 19th-century literature, Goldsmiths' Professor of English Literature at Oxford 1948–69.

[11] Sir Isaiah Berlin (1909–   ), Chichele Professor of Social and Political Theory at Oxford 1957–67.

[12] See 'Foreword' to *The Crowning Privilege: The Clark Lectures 1954–55* (1955), ix: 'The "crowning privilege" of the English poet is . . . his membership of a wholly anarchic profession. . . . His responsibility must be to the Muse alone.'

much his temper as his prose style. What is held against the two ladies is that they are in Oxford anyway: 'we could go to hear Enid or Helen any time we chose'. (Mr Griffith-Jones heard the latter lady to his cost in the Old Bailey not so long ago.)[13]

Mr W. W. Robson, Fellow of Lincoln College, who collected the signatures for Graves, quotes the precedent of Auden as a poet who is also an entertainer. A woman physiologist agrees: 'I definitely want an outsider and I definitely want a practising poet. The Faculty of English are concerned with the past; as a scientist I am concerned with the present—which presumably is also the case with poets.' In contrast to her are the dons who say: 'A vote for Helen Gardner is a blow for scholarly standards; when she lectures she uses a beginning, a middle and an end.' One wonders how Sir Henry Birkhead, who founded the Chair,[14] would vote; he was a poet—but only in Latin. And would he be shocked to find what dragon's teeth he has sown? I heard one don say of another: 'It is a case of the corpse nominating the killer.'

The undergraduates of course have no say in all this but then, as one don expressed it, 'the undergraduates are an eternal nuisance' (he was referring to their alleged current hankering for publicity). Still, it is the undergraduates who will benefit or suffer from a Professor of Poetry; it was they who got most pleasure and profit from the company of Auden as he held court in the Cadena over his morning coffee and it is they who stand to be most stimulated by the Old Scrutineer or the White Goddess. That either Graves or Leavis is likely to commit absurd generalizations and injustices in his lectures does not really matter. What undergraduates need is provocation. Auden's whinnying irreverences and *enfant terrible* eccentricities must have done them more good than crate-loads of scholarship.

So the new change in the regulations which, instead of confining the field to Oxford MAs, puts the Chair within the grasp of Tom, Dick and Harry—and of Graves and Leavis, is very much to be

[13] J. Mervyn G. Griffith-Jones (1909–79), the well-known British prosecutor at the trial of major war criminals at Nuremberg 1945–6, was Crown prosecutor in the 'trial of *Lady Chatterley's Lover*', Oct. 1960. For Dame Helen's testimony and Mr Griffith-Jones's shock see, *The Trial of Lady Chatterley: Regina v. Penguin Books Ltd.*, ed. 'C. H. Rolph' (pseud. of Cecil Rolph Hewitt) (1961), 58, 60.

[14] The endowment dates from 1696: John Ardagh, 'Socrates in the Cadena', *Observer*, loc. cit.

welcomed. The late Dylan Thomas for instance would now be eligible. As I was walking along Walton Street I remembered how W. R. Rodgers[15] and I were once stopped there at four o'clock in the morning by a policeman who was disappointed to find that our brief-cases did not contain knives and forks. When we told Thomas this next day he said: 'But why didn't you tell him you never travel with anything but a change of verse and a clean pair of rhymes?' This would be a good remark for a Professor, though I cannot hear it from the lips of either Dr Leavis or Miss Gardner—'the Lass with the Delegate Air' as they call her in the Clarendon Press.[16]

A typical snag in the proceedings occurred when it was discovered that both the local candidates were technically ineligible because they hold University Readerships, though the Chief Clerk in Registry remarked gently: 'Of course they are eligible—provided they resign their Readerships.' It must be remembered that the Chair of Poetry carries only £300 a year and, what seems deplorable, no perks. All was put right, however, by Congregation on Tuesday, so the four-cornered fight is still on. We can now take our four grains of salt and wait for the election on 16 February. Each of the contestants has some notabilities in his or her corner, but where Miss Gardner for instance has Sir Maurice Bowra, Graves has some more surprising seconds—Iris Murdoch, Kingsley Amis and John Wain.[17]

Oxford these days is acquiring a new face since the rotten Headington stone, much used in the 17th and 18th century buildings, including the Sheldonian Theatre itself, is being replaced by brash new stone from Derbyshire. Something symbolic here. Get out your sponges and towels, Amis and Wain, and don't let's have any more sugary talk of Lost Causes. Whoever wins this election, poetry itself should not lose; each of the four candidates would be a credit to the Chair.

Late at night in a gathering of academics I overheard someone say: 'If you can pronounce where you're going I'll drive you there.'

[15] W[illiam] R[obert] Rodgers (1909–69), Presbyterian minister turned poet, a BBC producer and script-writer, was a great Irish friend of MacNeice.

[16] Dame Helen was a delegate of the Oxford University Press 1959–75.

[17] Sir [Cecil] Maurice Bowra (1898–1971) had been Professor of Poetry at Oxford 1946–51. Iris Murdoch (1919–), Kingsley Amis, John Wain are British novelists. Wain was Professor of Poetry at Oxford 1973–8.

I should like to use this as a text for the election. Whether it is Arnold on Homer or Auden, as in his valedictory lecture, on Cooking, the Professor of Poetry must be a man who appreciates the words of others and can find his own words to convey this. He must know where his home or his heart is—*and* be able to pronounce it.

Pleasure in Reading: Woods to Get Lost In[1]

The Fishes and Fisheries of the Gold Coast[2] . . . Rearranging my books for the eighth or ninth time after a house-move and finding yet once more that in spite of losing the inevitable batch of favourites I have more volumes than I can accommodate, I think: surely *this* is one to scrap. But I open it and read: 'We are at present very ignorant of the biology of the Gold Coast fishes'[3] and then I remember watching the fishermen at Keta[4] (which is the end of the world) hauling on a mile-long horseshoe of seine-net and decide that some day, though God or Nkrumah[5] knows when, I may need this book for reference. In the same way I reprieve four volumes of Pareto's *The Mind and Society*,[6] a Greek translation of some Kipling short stories, *Le Sopha* (second volume only) by Crébillon *fils*[7] which I mistakenly bought when at school hoping to shock myself, three volumes inherited from my father of the *History of the Church of Ireland*,[8] Julian Huxley on

[1] *The Times*, no. 55162 (17 Aug. 1961), 11b–c.

[2] By Frederick Robert Irvine (1947).

[3] Ibid., 1 ('Bionomics').

[4] Seaport of East Ghana (Gold Coast), 85 miles ENE of Accra. MacNeice had visited the Gold Coast in Oct. 1956 for his radio feature 'The Birth of Ghana', broadcast 22 Feb. 1957.

[5] Kwame Nkrumah (1909–72), the first prime minister of Ghana and president of the republic, who liberated the Gold Coast from British colonial rule—the first British colony in Africa to become independent (18 Sept. 1956).

[6] By Vilfredo Pareto, trans. Andrew Bongiorno and Arthur Livingston, 4 vols. (1935), reviewed by MacNeice when it appeared: 'Charlatan-Chasing', *Morning Post*, 15 Nov. 1935, 6—to be published in the second volume of prose selections.

[7] Claude-Prosper Jolyot de Crébillon (1707–77), *Le Sopha* (1745; *The Sofa*), a satirical and licentious tale in an oriental setting.

[8] *History of the Church of Ireland from the Earliest Times to the Present Day*, ed. Walter Alison Phillips, 3 vols. (1933–4).

Evolution,[9] O. L. Owen on *Rugby Football*,[10] and the unreadable Latin plays of the medieval nun Hrotsvitha.[11] For it is pleasant to own books in the way a tit owns a coconut: it is there to peck at when you want it, otherwise the matter rests suspended.

There is an advantage in books which are primarily informative: since they need not be judged æsthetically, the reading of them need not be tainted by any hint of the snob game. Yet in this category there are many that I open for refreshment as well as for reference. To start with there are dictionaries and works of sheer statistics, *Wisden*[12] for example or *The Guinness Book of Records*; these can be great fun for anyone who enjoys the absurdities of language (who except the Germans would have named jewelry *Schmuck*?) or of fact (the dimensions of dinosaurs and chihuahuas). At the other end of the scale (in between come *The Mentality of Apes*[13] and the Abbé Dubois's *Hindu Manners, Customs and Ceremonies*)[14] are those works which, though not literature proper, have literary merit. One such of which I am very fond is Mayhew's *Studies of London Labour and the London Poor*.[15] This I have used as a case-book when digging for radio material, but far more often I have gone to it for its sheer entertainment value. There is perhaps some *Schadenfreude* involved in reading about all those orphans and mudlarks and pure-finders, but the appeal lies much more in the fact that

[9] *Evolution: The Modern Synthesis* (1942; 2nd edn 1963; 3rd edn 1974).

[10] Owen L. Owen, *The History of the Rugby Football Union* (1955).

[11] Hrotsvit (Roswitha) of Gandersheim, 10th-century Christian Saxon poët, imitated Terence in his Latin comedies. Her purpose 'was to oppose to Terence's representations of the frailty of women the chastity of Christian virgins and penitents': *The Oxford Dictionary of the Christian Church*, 2nd edn, ed. F. L. Cross and E. A. Livingstone (1974), 671–2.

[12] *John Wisden's Cricketers' Almanacks*.

[13] By Wolfgang Koehler, trans. from 2nd rev. edn by Ella Winter (1925; Penguin 1957; 1973).

[14] Jean Antoine Dubois, *Hindu Manners, Customs and Ceremonies*, trans. H. K. Beauchamp, 3 edns (1897, 1899, 1906 . . .). MacNeice was sent on BBC assignment to India and Pakistan for partition and independence in 1947.

[15] Henry Mayhew, 4 vols. (1861–2), a record of investigations into the life and work of the poor and underprivileged, using much of the street language of the people. There were many editions; Peter Quennell's edition *Mayhew's London* (1949) survived from MacNeice's library—one of his books now at the Butler Library, Columbia University. Cf. Dylan Thomas's delight in Mayhew as remarked by MacNeice in his review 'Round and About Milk Wood' (April 1954) above with n. 9.

Mayhew was a brilliant reporter with a very fine ear for natural speech.

The books which I most enjoy, of course, being directly or indirectly 'creative writing', force one to show one's hand, declare one's taste, get involved in argument. Still, since I am involved anyway, let me try to sort myself out. Of the works that I 'turn to time and again' the great majority are in verse. It is like having a taste for distilled liquors, natural in someone who is in the distillery business. Thus, with the exception of Dickens whom I shall return to, I rarely re-read the novels of the Old Masters. It seems to me that most novels are full of padding, of which we get quite enough in life. Of course most long poems have their share of padding too, but this is to some extent atoned for by the fact that they are in verse; in both reading and writing I happen to prefer verse to prose. All the same there are only a few really long poems that I am prepared to re-read *in toto* without feeling tired in advance. *Comus* yes, but *Paradise Lost* and *Paradise Regained* No! The easiest going for me is *The Faerie Queene*, not only because of its flow but because of its extraordinary variety; and Spenser of course knew far more about human beings than either Milton or Virgil. Another epic work which I home towards and linger in is Dante's *Inferno* (read with the Temple Classics crib) but *not* the rest of the *Divine Comedy*; neither Purgatory nor Paradise is rich enough in *story*. As for Homer, give me the *Odyssey* every time as against the *Iliad*.

With poetry I usually feel I know what the poets are trying to do but, as I have little notion what novelists are getting at, to read a novel is for me a blind date in the jungle, so the pleasure I get from it is often, presumably, fortuitous. Among the novels of our time *Ulysses* is a special case and I doubt if I should like it so much if I did not know the smell of Dublin. And with Virginia Woolf I re-read her for nostalgia and rhythm. Coming to the younger novelists I prefer such works as William Golding's *Lord of the Flies* which seem to exist, as poems do, on more than one plane;[16] in the same way among the newer playwrights I get far more pleasure from Beckett or Pinter than from Osborne, though

[16] *Lord of the Flies* (1954). 'Parable art' (Auden's coinage) was a literary issue addressed by MacNeice in *Varieties of Parable* (1965), his Clark Lectures at Cambridge 1963 which include comments on the English novelist William G. Golding (1911–   ).

when it comes to dialogue I fear I would usually rather hear it or write it than read it. But why is Dickens the only earlier novelist that I often re-read? Is it perhaps that he is an anti-novelist: that he is both too logical and too inconsequent, that his characters are far too rigid or over-simple to be characters while his props are too large and his settings too exuberant? I agree with George Orwell that 'the special Dickens atmosphere is created' by 'florid little squiggles on the edge of the page'[17] and I think it wicked of Robert Graves to have once attempted an 'Essential [i.e. potted] David Copperfield'.[18] Whether Flaubert or anyone else turns in his grave, this is an atmosphere I wallow in.

There are other kinds of prose work which come very high on my list of enjoyables. These could be placed very roughly in three groups: saga or legend, poetic fantasy, folk-tale. The first group has to be elastic if it is to include, say, the *Njal Saga*,[19] which was based on historic fact, is completely devoid of frills and in many ways anticipates the novel, but I would place it here because the movement of such sagas is that of Greek tragedy and also for the subjective reason that Ancient Iceland, though inhabited by hardheaded farmers and litigators, has entered the realm of legend through the sheer passage of time. But my classic example in this group would be Malory's *Morte D'Arthur*,[20] a wood which one can get lost in—but then what are woods for? Poetic fantasy, which is the opposite of idle whimsy, is represented by such diverse works as *The Golden Ass*,[21] *The Pilgrim's Progress*, the Chinese classic *Monkey* (translated by Arthur Waley),[22] and, inevitably, the *Alice* books. I might add John Wyndham's *The Day of the Triffids* which I only this year discovered to be bang

[17] Eric A. Blair (pseud. 'George Orwell', 1903–50), 'Charles Dickens' (1940), sect. v, in *The Collected Essays, Journalism and Letters*, ed. Sonia Orwell and Ian Angus, 4 vols. (1968), i, 451.

[18] *The Real David Copperfield*, 'by' (i.e. rewritten by) Robert Graves (1933); MacNeice's square brackets.

[19] *The Story of Burnt Njal*, trans. Sir George [Webbe] Dasent (1861, 1900, 1911). MacNeice had dramatized the *Njáls saga*, most famous of Icelandic sagas and a favourite story for MacNeice from at least his school days (*The Strings are False*, 98), in two radio adaptations broadcast 11–12 March 1947.

[20] See MacNeice's article 'Sir Thomas Malory' (April 1936) above.

[21] See 'Introduction to *The Golden Ass*' (1946) above. This work was also dramatized by MacNeice in two radio adaptations broadcast 3 and 7 Nov. 1944.

[22] Wu Ch'êng-ên (*c*. 1500–*c*. 1582), *Monkey* [abridgement of the *Hsi Yu Chi*], trans. Arthur Waley (1942).

on one of my own targets[23] (no characterization in it but that only makes it more direct). As for folk tales, I can lap them up from any country (they are often, indeed, the same tales) but still am fondest of the Norse ones. This after all is where we came in and we never needed Jung's permission to return there.

Blood and Fate[1]

The Odyssey. Translated by Robert Fitzgerald. (Heinemann. 35*s*.)
Patrocleia. Book XVI of Homer's *Iliad*. Adapted by Christopher Logue. (Scorpion Press. 15*s*.)

The first translation of the *Odyssey* that I read was by Butcher and Lang,[2] who are nowadays always good for a laugh with their 'grave house-dames' and other such quaint archaisms. All the same, Butcher and Lang have some of the main translator's virtues: unlike many of their rivals, such as T. E. Lawrence,[3] they are self-consistent; they do not keep making unnecessary additions and subtractions; although they write in prose, their sentences flow pleasantly; above all they carry you on. The current fashion in translation is to subtract rather than to add, leaving out stock epithets, proper names, and so on, either because they are no longer meaningful or because they do not conform to modern taste. Thus modern taste likes its diction to be penny plain. Does this mean that E. V. Rieu is the ideal translator of Homer?[4] One cannot see Homer agreeing; whether one likes it

[23] John Beynon Harris (pseud. 'John Wyndham', 1903–69), *The Day of the Triffids* (1951, Penguin 1954)—a science-fiction parable, tale of futurist terror, written 'on more than one plane' as MacNeice remarks above about Golding's *Lord of the Flies*. Commenting upon science fiction in *Varieties of Parable*, 26, MacNeice stated, 'a book like *The Day of the Triffids* seems to me closer to Spenser than to the main tradition of the English novel'.

[1] *Listener* 68: 1749 (4 Oct. 1962), 527.

[2] *The Odyssey of Homer, done into English prose*, trans. Samuel H. Butcher and Andrew Lang (1879), the well-known Victorian prose translation. Lang with Walter Leaf and Ernest Myers translated *The Iliad* (1883).

[3] 'T. E. Shaw' (pseud. of Thomas Edward Lawrence, 1888–1935), trans., *The Odyssey of Homer* (1932; 1955).

[4] Emil Victor Rieu (1887–1972), trans., *The Penguin Odyssey* (1945) and *The Penguin Iliad* (1950), both in plain prose.

or not, his diction is twopence coloured. To confine him to the ordinary language of twentieth-century men is as perverse as to rewrite Beatrix Potter in the way that was in fact suggested, i.e. where *her* lettuces were 'soporific', substituting lettuces that merely 'make you sleepy'. Similarly, repetition, whether of phrases or whole lines, is a basic characteristic of Homer as it is of all oral story-tellers; to argue that you are translating for the page is disingenuous, as the page *can* preserve a tone of voice. Such repetition may be primitive, but it corresponds to what George Orwell called the 'florid little squiggles on the edge of the page' that create 'the special Dickens atmosphere'.[5] Not but what there are people around prepared to rewrite Dickens too.

To tackle Homer at all needs courage and stamina. Of these two new books Christopher Logue's is the more original but it should not be called a 'translation'; anyhow it is taken from a single book of the *Iliad*. Robert Fitzgerald, known already both as a translator and a poet, has had a go at the whole *Odyssey*. He too carries you on, and he has the advantage over Butcher and Lang that he writes in verse, though I do not find he makes full use of it. I think it would have helped if he had attempted a line-by-line translation, keeping wherever possible the word order of the original; this kind of puzzle holds the translator up but, perhaps just because of that, a by-product can be that tautness which is so easily lost in translating a long poem. I also think he was mistaken in using a five-stress basically iambic line, though he often uses it felicitously:

> Then I addressed the blurred and breathless dead.[6]

And I do not think he succeeds with his occasional divagations into lyrical forms, as in the Song of the Sirens, any more than Robert Graves did in his translation of the *Iliad*.[7] Also some of the titles which he gives to the separate books are unfortunate: 'The Lord of the Western Approaches' suggests the second world war, while 'The Grace of the Witch' does not, to me at least,

[5] The same quotation is cited above: see 'Pleasure in Reading' (Aug. 1961) with n. 17.

[6] The American poet Robert [Stuart] Fitzgerald (1910–85), trans., Homer's *The Odyssey* (1962), xi. 166. Fitzgerald, a distinguished translator, became Boylston Professor of Rhetoric at Harvard 1965–84. He translated *The Iliad* (1974).

[7] *The Anger of Achilles: Homer's Iliad translated* (1959, 1960).

suggest Circe. Nor is he very happy with the stock-epithets: *polymetis* Odysseus, translated by Butcher and Lang as 'Odysseus of many devices', he renders sometimes 'the strategist Odysseus' and sometimes 'the great tactician', thereby losing the *earthiness* of his hero, just as 'the awesome one in pigtails' loses Athena's beauty. Still, as already indicated, he is readable—which is the first virtue in a translator—and his diction on the whole is clean and firm and often provides an edge where it is needed:

> How did you find your way down to the dark
> where these dimwitted dead are camped for ever,
> the after-images of used-up men?[8]

With Christopher Logue we are back to the world of Ezra Pound or further.[9] A few years ago Mr Logue contributed a version of the fight between Achilles and the river-god Scamander to a series of translations from the *Iliad* broadcast by the BBC;[10] his episode was outstanding both for strength and speakability. It was also far and away the freest, since Mr Logue knows no Greek at all. The same qualities are found in his *Patrocleia*. I spoke of avoiding additions and subtractions; this version is made up of both but then, as I said, it is *not* a translation. When Mr Logue wants to be succinct, he certainly is; at other times he embroiders on his original, as when he adds to the wasps in Homer's simile:

> *feeding their grubs*
> *And feeding off the sticky spit the grubs exude.*[11]

At other times he interpolates great lumps of Logue. This last practice sometimes makes for anachronism:

[8] Fitzgerald, trans., op. cit., xi. 180.

[9] i.e., spirited, free-verse adaptation in modern idioms rather than accurate translation, metrics, diction. See Pound's *Translations*, intro. Hugh Kenner (1953), and *Homage to Sextus Propertius* (1919, 1934) in *Personae: Collected Shorter Poems* (1952), 215–39.

[10] The English poet, playwright, and actor Christopher Logue's (1926–) translation of the *Iliad* xxi. 184–382 entitled 'The Battle with the River' was broadcast on the Third Programme 28 Feb. 1958.

[11] Christopher Logue, trans., *Patrocleia* : Book xvi of Homer's *Iliad* freely adapted (1962), 12; Logue's italics. Logue also 'adapted' Book xix of the *Iliad* as *Pax* (1967).

You know from books and talking pictures,
How people without firearms set about
Killing a tiger. . . . [12]

but on the whole this is a remarkable achievement of what it is fashionable to call 'empathy'. It brings out more vividly than most translations what both the Greeks and Trojans are up against: blood and sweat on the one hand and fate (the gods) on the other. Never was blood bloodier or fate more fatal. At the end of this tragic episode, Mr Logue, in Patroclus's dying speech, typically condenses and amplifies at once, even inventing a conceit which somehow seems legitimate:

I can hear Death
Calling my name and yet,
Somehow it sounds like '*Hector*'
And when I close my eyes
I see Achilles' face with Death's voice coming out of it.[13]

A Preface to 'The Faerie Queene'[1]

A Preface to 'The Faerie Queene'.
By Graham Hough (Duckworth. 25*s*.)

It is a serious reflection on literary taste to-day that so few people read *The Faerie Queene*. They should have been tempted to do so in 1936 when C. S. Lewis published *The Allegory of Love*, but here again they were probably frightened off that excellent book by its title; as if allegory were something as obsolete as chastity belts. So the non-readers of Spenser range from those who assume he is too dry, a dealer in dusty personified abstractions, a pedlar of verbal algebra, to those who assume he is (in the slang sense of the word) too 'wet', a hopelessly vague and repetitive dreamer who never grapples with realities. Professor Lewis could have shown both lots how wrong they are. So can Mr Graham Hough in this new book about our neglected Master. He begins by considering *The Faerie Queene* in relation to the

[12] *Patrocleia*, 31; Logue's italics. [13] Ibid., 35.

[1] *Listener* 69: 1766 (31 Jan. 1963), 213. Cf. MacNeice's comments on Spenser in his *Varieties of Parable*, chap. 2.

romantic Italian epic, but let that put nobody off. As he points out, this poem 'belongs to a forgotten kind', so the first thing to do, discarding 'the critical do-it-yourself kit supplied for use in colleges to-day',[2] is to establish what kind this is. In the process, Ariosto and Tasso are endeared to us.[3]

But this is far from being yet another thesis on literary influences. Without his Italian predecessors Spenser would never have written the way he did, and yet he is basically unlike them. Mr Hough is most helpful in diagnosing the Italianate element in Spenser, but he is more helpful still when he tries to pin down the Spenserian element. For Spenser, however much he borrowed—and sometimes he literally translates—remains enormously original. Mr Hough, like Professor Lewis before him, explodes many popular preconceptions. Spenser is not a mere dreamer or decorator, he is not overridingly a preacher, he is not continuously or narrowly allegorical, above all he is not monotonous. *The Faerie Queene* is an image of the world—'but' says Mr Hough, and it is an important 'but'—'an image of the interior world'.[4] If you are doped by a surfeit of 'realistic' novels you may not get the point but, if you have read Freud on dreams, or better still, dreamt yourself, you will know what Hough and Spenser are getting at. For here is to be found 'a huge panorama of man's inner experience, political, military, social, erotic, moral and religious'.[5]

Mr Hough's most original, if most difficult, chapter is on 'Allegory in *The Faerie Queene*'. He begins by tackling the well-worn but obscure distinction between 'allegory' and 'symbolism', behind which he spots 'a metaphysical spectre' lurking, and correctly, as against its detractors, states that 'Allegory in its broadest possible sense is a pervasive element in all literature.'[6] But, since broadest possible senses do not get one far in criticism, Mr Hough, using the two terms 'theme' and 'image', proceeds to work out, on the basis of the varying proportions of these two, a diagrammatic scheme which would cover the whole of literature. Sensibly, he makes this scheme circular: 'through one half of the circle we recede from simple allegory to the opposite extreme of

[2] Hough, *A Preface* (1962), 11.

[3] Ludovico Ariosto's *Orlando Furioso* (1532) and Torquato Tasso's *Gerusalemme Liberata* (1575), the famous Italian Renaissance epics, influenced Spenser.

[4] Hough, op. cit., 98. [5] Ibid. [6] Ibid., 102 and 105.

realism, while in the other half we return towards simple allegory again, by another route'.[7] If 'Naïve Allegory' is placed at twelve on the clock face, what Mr Hough calls 'Incarnation' (Shakespeare) will be found at three o' clock. Much of *The Faerie Queene* would come later than one o'clock, though none of it reaches 'Incarnation'.

It is a daring and fascinating scheme, and less unacceptable than most of its kind. It is certainly useful to Mr Hough in assessing the range—and also the limits—of *The Faerie Queene*, in which work, as he says, taken as a whole, 'allegory is not so decisive a factor, theme not so dominant over image, as we have sometimes been led to expect'.[8] By looking at the text itself—and he devotes a chapter to each of the six books and another to the Mutability Cantos—he easily proves his point that Spenser's allegory is sometimes naïve, often multivalent, and often, indeed, non-existent. As for the construction of this great poem, Mr Hough (I don't think I agree with him) does not regret that Spenser never finished it. Poets, as he points out, rarely know exactly what they are driving at and, whatever Spenser had *planned*, the poem that survives can be regarded as 'an organic growth with its own kind of wholeness. The legend of St George is the perfect beginning; Courtesy is the perfect end . . . Mutability is the perfect epilogue.'[9]

Yeats at Work[1]

Between the Lines. By Jon Stallworthy. (Oxford. 38*s*.)

Yeats was one of the slowest composers of poetry known to us: his 'stitching and unstitching', as he called it[2] made him a snail among Penelopes but, unlike Penelope, he always finished the job. He wrote to Lady Gregory: 'I, since I was seventeen, have never begun a story or poem or essay of any kind that I have not finished.'[3] And so, when the Variorum edition of his poems was

[7] Ibid., 106. [8] Ibid., 112. [9] Ibid., 235.

[1] *Listener* 69: 1773 (21 March 1963), 521.

[2] 'Adam's Curse' (1902), st. 1, also quoted at the beginning of his review of *The Variorum Edition of the Poems of W. B. Yeats* (Dec. 1958) above.

[3] Letter, 2 June 1900: *The Letters*, 345.

published in 1957, we thought: 'This is all very well, but what we now want to see are the work-sheets.'[4] This wish has been at last more than gratified by Mr Stallworthy, who, with great industry, intelligence, imagination, and taste, has collated the drafts of eighteen poems. These include 'The Second Coming', the haunting last poem 'The Black Tower', and, above all, the two Byzantium poems.

Some indication of Mr Stallworthy's industry—and Yeats's—is that for 'Coole Park, 1929', a poem of thirty-two lines, 'behind its comparatively simple structure and fluent rhythms lie thirty-eight pages of working'.[5] One could not often fault Mr Stallworthy's diagnoses of the poet's processes as revealed through these series of drafts. I find him on the whole, perhaps, just a shade too reverential (but that is a good fault), and occasionally naïve, as when he thinks the line 'Never, never will he come again'[6] must be an unconscious recollection of *King Lear*: has he never heard a child crying for a lost toy?

One must agree with Mr Stallworthy that Yeats's changes were nearly always improvements, though I still disagree about 'The Sorrow of Love', which, unlike the other poems here, was drastically rewritten after a long interval. The drafts reveal that 'The Second Coming' was originally meant to be in rhyme, that 'Chosen' and 'Parting' 'appear to have been conceived as parts of a much longer poem entitled "Two Voices"', and that 'Coole Park, 1929' was, to start with, more cluttered with the poet's ego—

A weathercock, shuttlecock, comeleon (*sic*).[7]

One of Mr Stallworthy's conclusions is that the elimination, or at least the restriction, of self was a characteristic of Yeats's creative procedure. The drafts bear this out, as they also bear out his other contentions that 'far from working from the informal to the formal as one might imagine, Yeats tended to free his rhythms as he advanced' and that 'an important and consistent feature of Yeats's revision is his tendency to cut the material with which he begins: seldom to add to it'.[8] It is also fascinating to

[4] See MacNeice's review (Dec. 1958) above.
[5] Stallworthy, 200.
[6] Ibid., 149. See *King Lear* V. iii. 308.
[7] Stallworthy, 140 and 184.
[8] Ibid., 246 and 251–2.

observe the cases of 'the word or phrase, which obviously pleases him, and which is often of central importance to the poem, that he is determined to work into its fabric somewhere'.[9] Yeats was exceptionally ruthless with himself, but he had his obsessions and compulsions.

One of Yeats's quaintest habits was to begin with a prose draft or sometimes a prose synopsis. These can be very revealing. Thus for 'Byzantium' he wrote in his diary: 'A walking mummy; flames at the street corners where the soul is purified. Birds of hammered gold singing in the golden trees. In the harbour (dolphins) offering their backs to the wailing dead that they may carry them to paradise. These subjects have been in my head for some time, especially the last.'[10] To be guided by Mr Stallworthy through the mazes and permutations of such material is like going on safari with a skilled hunter. But in 'Byzantium' he does make one interpretation that worries me. It is that mummy again. The final version reads:

> A mouth that has no moisture and no breath
> Breathless mouths may summon . . . [11]

'Breathless mouths' had been 'breathing mouths', and Mr Stallworthy explains: 'In both cases I think Yeats meant that *his* mouth—he was poised, for the purposes of the poem, between life and death—was able to summon his ghostly guide';[12] i.e. he makes Yeats the subject and the mummy the object. But the earlier drafts here printed, I would have thought, make it quite clear that it is the mummy that does the summoning—which is just what such a mummy would do. And it seems most unlikely that Yeats between one draft and another would have inverted the subject-object relationship.

But it is the great virtue of a book like this that the author himself supplies the ammunition with which we might shoot him down. Not that we would often want to.

[9] Ibid., 245.

[10] Ibid., 115; Yeats's parentheses. Cf. Yeats, 'Pages from a Diary written in 1930' (1944), iii, in *Explorations* (1962), 290.

[11] 'Byzantium' 1929 (1932), st. 2.

[12] Stallworthy, 132.

The Saga of Gisli[1]

The Saga of Gisli. Translated by George Johnston. (Dent. 21*s*.)

The Icelandic sagas are not to everyone's taste but, if you are one of those who relish their dry and subtle craft and the odd society they depict, you will certainly welcome this new translation and the accompanying notes and essay by Peter Foote. Mr Foote writes: 'The sagas are in the form of the novel, mixed narrative and dialogue, but they also accept some of the chief limitations of the drama: thoughts and emotions must be physically perceived, heard or seen in speech and action.'[2] This particular saga was written in the earlier part of the thirteenth century, incorporating some verses written about a generation earlier. Gisli was a real man who was outlawed about 960 and is said to have been an outlaw longer than any of the saga heroes except Grettir, the Samson of Iceland. But Grettir is something of an oaf, and Gisli is both a more complex and more sympathetic character. And, unlike Grettir's, his story shows, in Mr Foote's words, an unusual 'interplay and open conflict between personal and family honour and personal and family love'.[3]

On the other hand, in view of the fact that Gisli kills a man in his bed (admittedly by way of tit for tat), we may query whether *his* saga does, to quote Mr Foote again, exemplify 'the highest standards of fair play, touched by a certain magnanimity and even graciousness of mind'.[4] But after all, in spite of some surface similarities, we are not here in Dr Arnold's Rugby, and Gisli was outlawed forty years before Iceland accepted Christianity: Tit for Tat in that bleak landscape made more sense than Turning the Other Cheek. Anyhow, all true saga addicts will not only enjoy

[1] *Listener* 69: 1787 (27 June 1963), 1083. George B. Johnston (1913–), the translator and Canadian comic poet. See the ghost of Grettir in MacNeice's early poem 'Eclogue from Iceland' (1936) as well as 'Dark Age Glosses on the Grettir and Njal Sagas' (1960): *Collected Poems*, 40–7 and 484–5.

[2] Foote, 'An Essay on the Saga of Gisli and its Icelandic Background', *The Saga of Gisli*, trans. Johnston (1963), 105. Peter G. Foote was Professor of Scandinavian Studies at University College, London, 1963–83.

[3] Ibid., 106. [4] Ibid., 104.

this story but identify themselves with the hero. Early on, like old hands at woodcraft, they will recognize the small and scanty tracks that indicate trouble to come, such as the typical sinister understatement: 'There was never full friendliness between the brothers again.'[5] Or the laconic and fatalistic comment: 'It has gone as I thought it would; what we have just done will be of no use, and I think that fate will have its way over this.'[6] Much of the technique of this story-telling is, to use a stage word, 'throw-away'. The reader therefore must read watchfully.

The famous 'realism' of the sagas is qualified and complemented by two things: as regards content, by a supernatural element (though neither the tenth nor the thirteenth century would have found this unrealistic), and, as regards form, by the insertion into the straightforward prose narrative of a number of far from straightforward pieces of verse. The saga of Gisli contains its fair share of both these things. There is a highly efficient sorcerer (sorcery, of course, is merely his side-line, since like all good Icelanders his main business is farming) and, more originally, Gisli, when outlawed, has a series of dreams in which he is visited by a good dream woman and a bad dream woman (the latter, needless to say, has the greater impact). These dreams Gisli himself recounts in verse, thus linking our two non-realistic elements:

> I dreamt the doom goddess
> Draped my bushy straight-cut
> Hair with a dour headpiece,
> Hat all blood bespattered.[7]

Mr Johnston, in translating this peculiar brand of verse, has made something at least more acceptable than many of his predecessors. This is important because, whatever their value as poetry, these verse passages do, as he says, 'contain many of the keys of the story' (for example when 'Gisli spoke a verse which he should have kept to himself', this is overheard and precipitates his outlawing).[8] With the prose Mr Johnston is undeniably successful. He explains that, whereas his 'first intention was to make a twentieth-century telling of the saga', it later seemed to him 'that the "otherness" of the Icelandic was best preserved by

[5] *The Saga of Gisli*, ii. 3. [6] Ibid., vi. 8. [7] Ibid., xxxiii. 52.
[8] Ibid., xviii. 26.

letting its word order and idiom, especially in the shifting of tenses, play their part in the English too'.[9]

Frost[1]

The Poetry of Robert Frost. By Reuben Brower. (Oxford. 42*s.*)

It is over 10 years since Randall Jarrell wrote of Robert Frost: 'No poet has had even the range of his work more unforgivably underestimated by the influential critics of our time.'[2] This fault has been to some extent corrected since, but even now most of us tend to oversimplify this often simple-looking poet and try to squeeze him into far too narrow a pigeon-hole. He is not just a New England Edward Thomas: he has that underlying *dramatic* quality which makes a lyric more lyrical. The best lyric after all is a lyric *plus*, but this plusness is the hardest thing for a critic to analyse. Thus with Keats's Odes it is easy to say what the 'Nightingale' has got that the 'Ode to Autumn' has not got; but the reverse, which is at least as true, is a far more difficult thing to put one's finger on. What *has* the 'Ode to Autumn' got that the other Odes haven't? Answer: something extremely important. Professor Brower, in his most illuminating full-length study of Frost, brings in this Ode very aptly for purposes of comparison. 'Frost's poem', he writes, 'is symbolic in the manner of Keats's "To Autumn", where the over-meaning is equally vivid and equally unnamable.'[3] The reader may be surprised to learn that the poem of Frost's he has here been discussing is the apparently very 'easy' little anthology piece, 'Stopping by Woods on a Snowy Evening':

> My little horse must think it queer
> To stop without a farmhouse near
> Between the woods and frozen lake
> The darkest evening of the year.[4]

[9] Johnston, 'A Note on the Translation', xi.

[1] *New Statesman* 66: 1687 (12 July 1963), 46.

[2] Randall Jarrell, *Poetry and the Age* (1955), 62—a book reviewed by MacNeice (Sept. 1955): see above.

[3] Brower, 35. [4] (1923), second quatrain.

Jarrell went so far as to describe Frost as 'the standing, speaking reproach to any other good modern poet'.[5] Professor Brower's labour of love makes the reasons for such a statement explicit. I could wish that he had admitted as Jarrell does without mincing his words about it that there is a bad Frost as well as a good. However, the poems which he discusses with such penetration are mostly the product of the latter and one must be thankful for large mercies. Frost is a poet whom it is easy to read too fast: Professor Brower reads him slowly for us. He begins at the technical end, calling his first chapter 'Voice Ways'. Every day now we find critics who should know better making fools of themselves about versification. Thus several reviewers of Thom Gunn's last book[6] seemed to assume that his abandonment of recognizable rhythmical patterns for unrecognizable syllabic ones *automatically* marked an advance and betokened an increase in profundity. Similarly some idiot recently stated that rhyme in English poetry was now a thing of the past. So it is refreshing to find Professor Brower quoting Frost himself as saying: 'I had as soon write free verse as play tennis with the net down.'[7] Elsewhere, introducing his *Collected Poems*, Frost wrote: 'The possibilities for tune from the dramatic tones of meaning struck across the rigidity of a limited metre are endless.'[8]

I have often been surprised that reviewers of verse pay so little attention to syntax. A sentence in prose is struck forward like a golf ball; a sentence in verse can be treated like a ball in a squash court. Frost, as Brower brings out, is a master of angles: he quotes Edwin Muir on his poetic 'method': 'starting from a perfectly simple position we reach one we could never have foreseen.'[9] The book throws a good deal of light on this process but, above all, it sends us back to the poems which are the source of that light. 'Hyla Brook', 'West Running Brook', 'For Once, Then, Something', 'Two Look at Two' . . . these and many more Brower treats individually, and always without skimping: Frost emerges as a master organizer. And Brower makes helpful comparisons with both Wordsworth and Emerson. Emerson, of course, as a poet (how could anyone appear so facile and uncouth

[5] Jarrell, op. cit., 36. [6] *My Sad Captains and Other Poems* (1961).
[7] Brower, 6. [8] 'The Figure a Poem Makes' (1939), para. 2.
[9] In *Recognition of Robert Frost*, ed. Richard Thornton (1937), 311, quoted by Brower, 171.

simultaneously?), is out of his class here, but his ideas of Nature were something in Frost's background, which also included those of Thoreau. It is Wordsworth who is most relevant but the differences are as important as the similarities. Brower writes of Wordsworth's 'To the Cuckoo': 'The transforming inner ear almost transforms the bird out of existence.'[10] The birds in the best Frost stay obstinately put in existence. And Frost is a more dramatic poet than Wordsworth, underlining conflicts where Wordsworth's 'art lies in blurring the differences'.[11] He also, as Brower emphasizes, has a very great measure of something most un-Wordsworthian—irony. T. E. Hulme, it will be remembered, cried out for a revival of irony.[12] Frost has revived it as much as any.

Professor Brower regards as 'a culminating poem in Frost's career' *A Masque of Reason*, which Jarrell, correctly I think, considers 'a frivolous, trivial and bewilderingly corny affair'.[13] And Jarrell has complained, in general, that the bad, and mainly elder, Frost is 'full of complacent wisdom and cast-iron whimsy'.[14] The odd thing is that the good Frost is anything but complacent and can be one of the most sinister writers in the language. Read the terrifying little poem, 'Design':

> What but design of darkness to appal?[15]

Again there is the poem where, thinking of Pascal,[16] he writes:

> I have it in me so much nearer home
> To scare myself with my own desert places.[17]

Frost knows loneliness only too well. But he also, as is made abundantly clear in his eclogues, knows human beings—in their own right and not as projections of himself. It is this combination of two qualities not often combined in modern poetry that gives him his commanding stature. Together with his very subtle craftsmanship. He has often been compared with Hardy but I find him

[10] Brower, 45. [11] Ibid., 76.

[12] Hulme, 'Romanticism and Classicism', in *Speculations*, ed. Herbert Read, 2nd edn. (1936), 113–40.

[13] Jarrell, 41. [14] Ibid., 40. [15] 'Design' (1936), l. 13.

[16] *'Le coeur a ses raisons que la raison ne connaît point'*: 'The heart has its reasons which reason knows nothing of': Pascal, *Pensées* iv. 277.

[17] 'Desert Places' (1934), st. 4.

a far more professional poet. We should be grateful to Professor Brower for such a professional study. And I must add that he does it without using jargon.

Louis MacNeice Writes . . . [on *The Burning Perch*][1]

When I assembled the poems in *The Burning Perch* (I am not happy about the title but could not think of anything better), I was taken aback by the high proportion of sombre pieces, ranging from bleak observations to thumbnail nightmares. The proportion is far higher than in my last book, *Solstices*, but I am not sure why this should be so. Fear and resentment seem here to be serving me in the same way as Yeats in his old age claimed to be served by 'lust and rage',[2] and yet I had been equally fearful and resentful of the world we live in when I was writing *Solstices*. All I can say is that I did not set out to write this kind of poem: they happened. I am reminded of Mr Eliot's remark that the poet is concerned not only with beauty but with 'the boredom and the horror and the glory'.[3] In some of the poems in *The Burning Perch* the boredom and the horror were impinging very strongly, e.g. the former in 'Another Cold May' or 'October in Bloomsbury' and the latter in 'Flower Show', 'After the Crash', 'Charon' or 'Budgie'. I find, however, that in most of these poems the grim elements are mixed with others, just as there are hardly any examples of pure satire in this collection; 'This is the Life', I suppose, comes nearest to it but still seems to me no more purely satirical than, for example, a medieval gargoyle.

[1] *Poetry Book Society Bulletin*, no. 38 (Sept. 1963), [1]. MacNeice's *The Burning Perch* (Faber, 13 Sept. 1963) was the Autumn 1963 Choice of the Poetry Book Society. Editorial note: 'This contribution by Louis MacNeice must have been one of the last things he wrote before his death on 3 September 1963. He sent it with a letter dated August 26th, apologising for delay and saying "my doctor won't let me go to London yet, so everything is awkward." '

[2] Yeats, 'The Spur' (1938–9), l. 1.

[3] 'But the essential advantage for a poet is not, to have a beautiful world with which to deal: it is to be able to see beneath both beauty and ugliness; to see the boredom, and the horror, and the glory.'—T. S. Eliot, 'Matthew Arnold', *The Use of Poetry and the Use of Criticism* (1933), 106.

When I say that these poems 'happened', I mean among other things that they found their own form. By this I do not, of course, mean that the form was uncontrolled: some poems chose fairly rigid patterns and some poems loose ones but, once a poem had chosen its form, I naturally worked to mould it to it. Thus, while I shall always be fond of rhyme and am sorry for those simple-minded people who proclaim that it is now outmoded (after all it remains unbeatable for purposes of epigram), a good third of the poems in this book are completely without it. Similarly with rhythm: I notice that many of the poems here have been trying to get out of the 'iambic' groove which we were all born into. In 'Memoranda to Horace' there is a conscious attempt to suggest Horatian rhythms (in English of course one cannot do more than suggest them) combined with the merest reminiscence of Horatian syntax. This technical Horatianizing appears in some other poems too where, I suppose, it goes with something of a Horatian resignation. But my resignation, as I was not brought up a pagan, is more of a fraud than Horace's: 'Memoranda to Horace' itself, I hope, shows this. So here again, as in poems I was writing thirty years ago (I myself can see both the continuity and the difference), there are dialectic, oxymoron, irony. I would venture the generalization that most of these poems are two-way affairs or at least spiral ones: even in the most evil picture the good things, like the sea in one of these poems, are still there round the corner.

A Bibliography of Short Prose by Louis MacNeice

(The items from the *Marlburian* were verified by John Hilton and Anthony Blunt.)

'The Story of the "Great Triobal Clan" as Created by Mr Schinabel', *Marlburian* 59: 854 (23 Oct. 1924), 135–6.

'The Four Ends of the World', *Marlburian* 60: 860 (28 May 1925), 65–6.

'The Devil' [Mr Schinabel, Part 2], *Marlburian* 60: 863 (22 Oct. 1925), 117–18.

'Windows in the Ink' [Mr Schinabel, Part 3], *Marlburian* 61: 867 (10 March 1926), 34–5.

'Jam of Aganippe', *Marlburian* 61: 868 (29 March 1926), 54.

'Apollo on the Old Bath Road (A School Serial)' I–II, *Marlburian* 61: 869 (26 May 1926), 65.

'Miss Ambergris & King Perhaps', *Marlburian* 61: 869 (26 May 1926), 66–7.

'Apollo on the Old Bath Road' II–V, *Marlburian* 61: 870 (23 June 1926), 90–1.

'Apollo on the Old Bath Road' V–X, *Marlburian* 61: 871 (26 July 1926), 118–19.

'Fires of Hell', *Cherwell*, NS, 18: 7 (4 Dec. 1926), 251.

Collected Poems, by James Stephens; *Transition*, by Edwin Muir, *Cherwell* 18: 8 (11 Dec. 1926), 329.

Autobiographies, by W. B. Yeats; *Human Bits*, by Hildegarde Hume Hamilton, *Cherwell*, 19: 1 (29 Jan. 1927), 28.

Christopher Marlowe, by U. M. Ellis Fermor, *Cherwell* 19: 4 (19 Feb. 1927), 129.

A Drunk Man Looks at the Thistle, by Hugh MacDiarmid, *Cherwell* 19: 6 (5 March 1927), 168.

Duncan Dewar's Accounts: A Student of St Andrew's 100 Years Ago, *Cherwell* 20: 3 (21 May 1927), 88.

'The Front Door of Ivory', *Cherwell* 20: 6 (11 June 1927), 185–6.

'American Universities (As gathered from *The Michigan Daily*, 8 pages, price 5 cents)', *Cherwell* 20: 8 (25 June 1927), 230.

The Enormous Room, by E. E. Cummings, *Oxford Outlook* 10: 47 (Nov. 1928), 171–3.

The Venture, edited by Anthony Blunt, H. Romilly Fedden and Michael Redgrave: No. 1, Cambridge, Nov. 1928, *Oxford Outlook*, 10: 47 (Nov. 1928), 184–5.

The Trial of Socrates, by Coleman Phillipson, *University News* (Oxford), 1: 7 (1 Dec. 1928), 253.

'Garlon and Galahad', *Sir Galahad* 1: 1 (21 Feb. 1929), 15.

'Foreword' to his *Blind Fireworks* (London: Gollancz, March 1929), 5–6.

'Our God Bogus', *Sir Galahad* 2 (14 May 1929), 3–4.

Poems, by W. H. Auden, *Oxford Outlook* 11: 52 (March 1931), 59–61.

The Lysis, translated by K. A. Matthews and illustrated by Lynton Lamb, *Oxford Outlook* 11: 55 (June 1931), 146–7.

Deserted House—a Poem-Sequence, by Dorothy Wellesley; *Dear Judas*, by Robinson Jeffers; *This Experimental Life*, by Royall Snow; *The Signature of Pain*, by Alan Porter, *Oxford Outlook* 11: 55 (June 1931), 147–9.

'Miss Riding's "Death"' (*The Life of the Dead*, by Laura Riding), *New Verse*, no. 6 (Dec. 1933), 18–20.

'Poems by Edwin Muir' (*Variations on a Time Theme*, by Edwin Muir), *New Verse*, no. 9 (June 1934), 18, 20.

Reply to 'An Enquiry', *New Verse*, no. 11 (Oct. 1934), 2, 7.

The Domain of Selfhood, by R. V. Feldman, *Criterion* 14: 54 (Oct. 1934), 160–3.

'Greek Classical Writers' (*A History of Classical Greek Literature*, by T. A. Sinclair), *Spectator* 154: 5559 (11 Jan. 1935), 58–9.

'Plato on Knowledge' (*Plato's Theory of Knowledge*, by F. M. Cornford), *Spectator* 154: 5571 (5 April 1935), 575–6.

'A Comment' [on G. M. Hopkins], *New Verse*, no. 14 (Hopkins Issue: April 1935), 26–7.

'A Brief for Cicero' (*Cicero: A Study*, by G. C. Richards), *Spectator* 154: 5574 (26 April 1935), 700.

'Modern Writers and Beliefs' (*The Destructive Element*, by Stephen Spender), *Listener* 13: 330 (8 May 1935), suppl. xiv.

'Translating Aeschylus' (Aeschylus: *The Seven Against Thebes*, translated by Gilbert Murray), *Spectator* 154: 5576 (10 May 1935), 794.

'Plato Made Easy' (*The Argument of Plato*, by F. H. Anderson), *Spectator* 155: 5584 (5 July 1935), 28.

'Fata Vocant' (*Religion in Virgil*, by Cyril Bailey), *Spectator* 155: 5589 (9 Aug. 1935), 231–2.

'Mr Empson as a Poet' (*Poems*, by William Empson), *New Verse*, no. 16 (Aug.–Sept. 1935), 17–18.

'Poetry To-day', *The Arts To-Day*, ed. Geoffrey Grigson (6 Sept. 1935), 25–67.

'On All Souls' Night' (*The Dark Glass*, by March Cost; *The Lion Beats the Unicorn*, by Norah C. James; *Don Segundo Sombra*, by Ricardo Guiraldes; *Richard Savage*, by Gwyn Jones), *Morning Post*, 10 Sept. 1935, 14.

'Town and Country in America' (*Summer in Williamsburg*, by Daniel Fuchs; *He Sent Forth a Raven*, by Elizabeth Madox Roberts; *Two Walk Together*, by Barbara Cooper; *The Collected Ghost Stories of Oliver Onions*), *Morning Post*, 17 Sept. 1935, 14.

'The Russian Mind and Plato: A Curious Blend' (*Plato*, by Vladimir Solovyev), *Morning Post*, 20 Sept. 1935, 14.

'A Story of the "Wrong Generation"' (*A Life's Journey*, by Hermynia Zur Muhlen; *Idle Hands*, by Edward Charles; *Derby and Joan*, by Maurice Baring; *Blow for Balloons*, by W. J. Turner), *Morning Post*, 24 Sept. 1935, 14.

'The Modern Short Story: Too Much Restraint, Too Little Verve' (*The Best Short Stories, 1935*, edited by E. J. O'Brien; *Moving Pageant*, by L. A. Pavey; *Round Up*, by Ring W. Lardner; *15 Odd Stories*, by Shane Leslie), *Morning Post*, 1 Oct. 1935, 14.

'Strange Tale of a Bad Hat: Mr Thompson's Novel of Nonconformity' (*Nottke the Thief*, by Sholem Asch; *It Cannot be Stormed*, by Ernest von Salomon; *Introducing the Arnisons*, by Edward Thompson; *Odd John*, by Olaf Stapledon), *Morning Post*, 8 Oct. 1935, 14.

'Mr David Garnett's Tale of a Tramp' (*Beaney-Eye*, by David Garnett; *River Niger*, by Simon Jesty; *Portrait of the Bride*, by Betty Miller; *On Approval*, by Dorothy Whipple), *Morning Post*, 15 Oct. 1935, 16.

'Another Tolstoy and His Fine Novel' (*Darkness and Dawn*, by Alexei Tolstoy; *Life Begins*, by Christa Winsloe; *Tomorrow is Also a Day*, by Romilly Cavan; *Hallelujah Chorus*, by Chris Massie), *Morning Post*, 29 Oct. 1935, 14.

'Sinclair Lewis Imagines Fascism in America' (*It Can't Happen Here*, by Sinclair Lewis; *The Asiatics*, by Frederic Prokosch; *Once We Had a Child*, by Hans Fallada), *Morning Post*, 5 Nov. 1935, 16.

'"The Green Child" and Another Poet's Novel' (*The Green Child*, by Herbert Read; *King Coffin*, by Conrad Aiken; *Girl of Good Family*, by Lucian Wainwright; *Cut and Come Again*, by H. E. Bates), *Morning Post*, 12 Nov. 1935, 14.

'Charlatan-Chasing: How We Live and How We Think: A World-Famous Book Translated' (*The Mind and Society*, by Vilfredo Pareto: edited by Arthur Livingston), *Morning Post*, 15 Nov. 1935, 6.

'The American World, and a Tale from Trinidad' (*Eclipse*, by Dalton Trumbo; *The Sun Sets in the West*, by Myron Brinig; *Quagmire*, by Henry K. Marks; *Black Fauns*, by Alfred H. Mendes), *Morning Post*, 19 Nov. 1935, 15.

'A Vast Novel Which Is Not a Bore' (*Summer Time Ends*, by John Hargrave; *Deep Dark River*, by Robert Ryles; *Out for a Million*, by V. Krymov; *Dust Over the Ruins*, by Helen Ashton), *Morning Post*, 26 Nov. 1935, 14.

'Ireland—a Tragic, Realistic Presentation' (*Holy Ireland*, by Norah Hoult; *Till She Stoop*, by Morna Stuart; *Swami and Friends*, by R. J. Narayan; *Tortilla Flat*, by John Steinbeck), *Morning Post*, 3 Dec. 1935, 14.

'Plato and Platonists' (*Plato's Thought*, by G. M. A. Grube; *Plato*, by Vladimir Solovyev), *Spectator* 155: 5608 (20 Dec. 1935), 1037–8.

'Some Notes on Mr Yeats' Plays', *New Verse*, no. 18 (Dec. 1935), 7–9.

Letter, in reply to St John Ervine's articles 'Our Peevish Poets' [in preceding issues], *Observer*, no. 7549 (2 Feb. 1936), 13.

The Achievement of T. S. Eliot, by F. O. Mathiessen [unsigned review], *Listener* 15: 372 (26 Feb. 1936), 414.

'The Newest Yeats' (*A Full Moon in March*, by W. B. Yeats), *New Verse*, no. 19 (Feb.–March 1936), 16.

'Sir Thomas Malory', *The English Novelists*, ed. Derek Verschoyle (30 April 1936), 17–28.

'The Passionate Unbeliever' (*Lucretius, Poet and Philosopher*, by E. E. Sikes), *Spectator* 156: 5628 (8 May 1936), 846.

'Fiction' (*I'd Do It Again*, by Frank Tilsley; *The Queen's Doctor*, by Robert Neumann; *Overture, Beginners!*, by John Moore; *A Week by the Sea*, by Bryan Guiness; *To the Mountain*, by Bradford Smith; *Please Don't Smile*, by Johann Rabener), *Spectator* 156: 5631 (29 May 1936), 994.

Collected Poems, 1909–1935, by T. S. Eliot [unsigned review], *Listener* 15: 388 (17 June 1936), 1175.

Longinus on the Sublime, by Frank Granger, *Criterion* 15: 16 (July 1936), 697–9.

'Fiction' (*Farewell Romance*, by Gilbert Frankau; *Standing Room Only*, by Walter Greenwood; *Fifty Roads to Town*, by Frederick Nebel; *Spring Storm*, by Alvin Johnson; *Rising Tide*, by Elisaveta Fen), *Spectator* 157: 5639 (24 July 1936), 156.

Dramatis Personae, by W. B. Yeats, *Criterion* 16: 62 (Oct. 1936), 120–2.

'Preface', *The Agamemnon of Aeschylus*, trans. MacNeice (Oct. 1936), 7–9.

'Translator's Note' [on production of his Aeschylus' *Agamemnon*], theatre programme, Westminster Theatre, 1 and 8 Nov. 1936.

Till the Cows Come Home, Geoffrey Kerr, *Time and Tide* 17: 45 (7 Nov. 1936), 1562.

'The "Agamemnon" of Aeschylus' [letter, in reply to reviewer of Group Theatre production], *The Times*, 12 Nov. 1936, 10c.

Letter [on review of *Agamemnon* production on 7 Nov.], *Time and Tide*, 17: 47 (21 Nov. 1936, 1632.

'Extracts from a Dialogue on the Necessity for an Active Tradition and Experiment' [by Louis MacNeice and Rupert Doone], *Group Theatre Paper*, no. 6 (Dec. 1936), 3–5.

'Letter from Louis MacNeice', *Cherwell* 48:8 (5 Dec. 1936), 190.

Look, Stranger! Poems, by W. H. Auden [unsigned review], *Listener* 16: 416 (30 Dec. 1936), 1257.

'Subject in Modern Poetry', *Essays and Studies* by Members of the English Association, coll. Helen Darbishire, 22, [Dec.] 1936 (1937), 144–58.

'A Parable' [concl. to 'A Dialogue' in preceding number], *Group Theatre Paper*, no. 7 (Jan. 1937), 2–3.

'Pax Romana' (*The Cambridge Ancient History*, volume xi: *The Imperial Peace*), *Spectator* 158: (8 Jan. 1937), 54–5.

'Jean Cocteau in English' (*The Infernal Machine*, by Jean Cocteau: English Version and Introductory Essay by Carl Wildman), *London Mercury* 35: 208 (Feb. 1937), 430–1.

'Fiction' (*Invasion '14*, by Maxence van der Meersch: translated by Gerard Hopkins; *Three Bags Full*, by Roger Burlinghame; *The Sisters*, by Myron Brinig; *The Porch*, by Richard Church; *Olive E*, by C. H. B. Kitchin; *Women Also Dream*, by Ethel Mannin; *Spring Horizon*, by T. C. Murray), *Spectator* 158: 5669 (19 Feb. 1937), 330.

'Fiction' (*Devil Take the Hindmost*, by Frank Tilsley; *One Life, One Kopeck*, by Walter Duranty; *The Other Side*, by Stephen Hudson; *The Moon in the South*, by Carl Zuckmayer; *Hallelujah, I'm a Bum*, by Louis Paul; *Perilous Sanctuary*, by D. J. Hall), *Spectator* 158: 5671 (5 March 1937), 422.

'Fiction' (*Very Heaven*, by Richard Aldington; *Time to be Going*, by R. H. Mottram; *The Dance Goes On*, by Louis Golding; *The Picnic*, by Martin Boyd; *At Last We Are Alone*, by F. Sladen-Smith; *The Tomato-Field*, by Stuart Engstrand), *Spectator* 158: 5673 (19 March 1937), 550, 552.

'Fiction' (*A Trojan Ending*, by Laura Riding; *Spanish Fire*, by Hermann Kesten; *Golden Peacock*, by Gertrude Atherton; *Maiden Castle*, by John Cowper Powys; *A Bridge to Divide Them*, by Goronwy Rees; *The Paradoxes of Mr Pond*, by G. K. Chesterton), *Spectator* 158: 5675 (2 April 1937), 632.

'The Bradfield Greek Play: *Oedipus Tyrannus*', *Spectator* 158: 5687 (25 June 1937), 1187.

The Note-Books and Papers of Gerard Manley Hopkins, edited with Notes and a Preface by Humphry House, *Criterion*, 16: 65 (July 1937), 698–700.

The Disappearing Castle, by Charles Madge; *Poems*, by Rex Warner; *The Fifth Decad of Cantos*, by Ezra Pound [unsigned review], *Listener* 18: 451 (1 Sept. 1937), 467.

'The Hebrides: A Tripper's Commentary', *Listener* 18: 456 (6 Oct. 1937), 718–20.

Lament for the Death of a Bullfighter and Other Poems, by Federico Garcia Lorca: translated by A. L. Lloyd [unsigned review], *Listener* 18: 456 (6 Oct. 1937), 744, 747.

'Chinese Poems' (*The Book of Songs*, translated from the Chinese by Arthur Waley), *Listener* 18: 457 (13 Oct. 1937), suppl. viii.

'With the Kennel Club', *Night and Day* 1: 16 (14 Oct. 1937), 40.

The Joy of It, by Littleton Powys [unsigned review], *Listener* 18: 462 (17 Nov. 1937), 1094.

'Letter to W. H. Auden', *New Verse*, nos. 26–7 (Auden Double Issue: Nov. 1937), 11–13.

Brief statement, in *Authors take sides on the Spanish War* (London: Left Review, Dec. 1937), unpaginated.

'In Defence of Vulgarity', *Listener* 18: 468 (29 Dec. 1937), 1407–8.

'The Play and the Audience', *Footnotes to the Theatre*, ed. R. D. Charques (1938), 32–43.

'Letter to the Editor', *New Verse*, no. 28 (Jan. 1938), 18.

The Best Poems of 1937, selected by Thomas Moult; *New Oxford Poetry 1937*, edited by Nevill Coghill and Alistair Sandford [unsigned review], *Listener* 19: 469 (5 Jan. 1938), 45–6.

It's Perfectly True and Other Stories, by Hans Christian Andersen [unsigned review], *Listener* 19: 472 (26 Jan. 1938), 207–8.

No More Peace, by Ernst Toller; *The Fall of the City*, by Archibald MacLeish; *Four Soviet Plays* [unsigned review], *Listener* 19: 474 (9 Feb. 1938), 321.

'A Statement', *New Verse*, nos 31–2 (Autumn 1938), 7.

The Oxford Book of Light Verse, chosen by W. H. Auden [unsigned review], *Listener* 20: 514 (17 Nov. 1938), 1079.

On the Frontier, by W. H. Auden and Christopher Isherwood, *Spectator*, 161: 5760 (18 Nov. 1938), 858.

'A Brilliant Puritan' (*Collected Poems*, by Robert Graves; *Collected Poems of Hart Crane*; *Tender Only to One*, by Stevie Smith), *Listener* 20: 517 (8 Dec. 1938), suppl., viii.

'Today in Barcelona', *Spectator* 162: 5769 (20 Jan. 1939), 84–5.

'The *Antigone* of Sophocles', *Spectator* 162: 5776 (10 March 1939), 404.

[Prefatory] Note, March 1939, to *Autumn Journal* 1938 (May 1939), in *Collected Poems*, ed. E. R. Dodds (1966, 1979), 101.

Letter 'Painters and Poets' [on William Coldstream], *New Verse*, NS, 1: 2 (May 1939), 58–61.

'Original Sin' (*The Family Reunion*, by T. S. Eliot), *New Republic* 98: 1274 (3 May 1939), 384–5.

'Four Contemporary Poets' (*Dead Reckoning*, by Kenneth Fearing; *M.1000 Autobiographical Sonnets* by Merrill Moore; *In Dreams*

Begin Responsibilities, by Delmore Schwartz; *Mirrors of Venus: Sonnets*, by John Wheelwright), *Common Sense* 8: 6 (June 1939), 23–4.

'Tendencies in Modern Poetry: Discussion between F. R. Higgins and Louis MacNeice, broadcast from Northern Ireland', *Listener* 22: 550 (27 July 1939), 185–6.

Letter, in reply to Peter Fleming's article 'Pax Bloomsburiana' in preceding issue, *Spectator* 168: 5811 (10 Nov. 1939), 652.

'A Bran Tub' (*The Christmas Companion*, edited by John Hadfield), *Listener* 22: 569 (7 Dec. 1939), suppl., xv.

'The Poet in England Today', *New Republic* 102: 13 (25 March 1940), 412–13.

'Not Tabloided in Slogans' (*Another Time: Poems*, by W. H. Auden), *Common Sense* 9: 4 (April 1940), 24–5.

'Housman in Retrospect' (*The Collected Poems of A. E. Housman*), *New Republic* 102: 18 (29 April 1940), 583.

'Yeats's Epitaph' (*Last Poems and Plays*, by W. B. Yeats), *New Republic* 102: 26 (24 June 1940), 862–3.

'American Letter' [to Stephen Spender], *Horizon* 1: 7 (July 1940), 462, 464.

'Oxford in the Twenties', *Partisan Review* 7: 6 (Nov.–Dec. 1940), 430–9.

'John Keats', *Fifteen Poets: Chaucer, Spenser, Shakespeare, Milton* [*and others*] . . . (Oxford: The Clarendon Press, Jan. 1941), 351–4.

'Scottish Poetry' (*A Golden Treasury of Scottish Poetry*, selected and edited by Hugh MacDiarmid), *New Statesman and Nation*, NS, 21: 517 (18 Jan. 1941), 66.

'London Letter: Blackouts, Bureaucracy & Courage', 1 Jan., *Common Sense* 10: 2 (Feb. 1941), 46–7.

Freedom Radio [film review], *Spectator* 166: 5875 (31 Jan. 1941), 116.

Quiet Wedding—The Mark of Zorro [film review], *Spectator* 166: 5876 (7 Feb. 1941), 143.

'Through Stained Glass' (*Some Memories of W. B. Yeats*, by John Masefield), *Spectator* 166: 5876 (7 Feb. 1941), 152.

Tin Pan Alley—Arise my Love—Arizona [film review], *Spectator* 166: 5877 (14 Feb. 1941), 172.

'Acknowledgement' and 'Foreword' to *Poems 1925–1940* (17 Feb. 1941), v and xiii.

'Traveller's Return', *Horizon* 3: 14 (Feb. 1941), 110–17.

'London Letter: Anti-defeatism of the Man in the Street', 1 March, *Common Sense* 10: 4 (April 1941), 110–11.

'Touching America', *Horizon* 3: 15 (March 1941), 207–12.

Angels Over Broadway—The Trail of the Vigilantes [film review], *Spectator* 166: 5882 (21 March 1941), 307.

Spare a Copper—The Ghost Train [film review], *Spectator* 166: 5883 (28 March 1941), 343.

'Acknowledgements', *Plant and Phantom* (April 1941), 11.
Victory—Mr and Mrs Smith [film review], *Spectator* 166: 5884 (4 April 1941), 371.
'The Way We Live Now: IV', *Penguin New Writing*, no. 5 (April 1941), 9–14.
'Cook's Tour of the London Subways' (radio feature condensed), *Listener* 25: 640, (17 April 1941), 554–5.
'London Letter: War Aims; the New Political Alignment', 1 April, *Common Sense* 10: 5 (May 1941), 142–3.
'The Tower that Once', *Folios of New Writing* 3 (Spring 1941), 37–41.
'The Morning after the Blitz', *Picture Post* 2: 5 (3 May 1941), 9–12, 14.
'U.S.A.' (*Who are the Americans?* by William Dwight Whitney), *New Statesman and Nation* 21: 532 (3 May 1941), 466–7.
'Autobiographies' (*In Search of Complications*, by Eugene de Savitsch; *King's Messenger, 1918–1940*, by George P. Antrobus; *Pioneering Days*, by Thomas Bell), *New Statesman and Nation* 21: 534 (17 May 1941), 512–13.
'London Letter: Democracy versus Reaction & Luftwaffe', 1 May, *Common Sense*, 10: 6 (June 1941), 174–5.
'London Letter: Reflections from the Dome of St Paul's', 2 June, *Common Sense* 10: 7 (July 1941), 206–7.
'"The Third Christmas"', *Radio Times* 73: 951 (19 Dec. 1941), 7.
Autobiographical sketch, in *Twentieth Century Authors: A Biographical Dictionary of Modern Literature*, ed. Stanley J. Kunitz and Howard Haycraft (1942), 888–9.
'Our Fourth Wartime Christmas', *London Calling*, 18 Dec. 1942, 4–5.
'Introduction: Some Comments on Radio Drama', *Christopher Columbus* (March 1944), 7–19.
'Oddity of "The Golden Ass"' [comment on his adaptations of Apuleius' *The Golden Ass*, *Cupid and Psyche*], *Radio Times* 85: 1100 (27 Oct. 1944), 5.
'Note', *Springboard: Poems 1941–1944* (Dec. 1944), 7.
'"The Golden Ass" or "Metamorphoses" of Apuleius: adapted by Louis MacNeice [comment on the rebroadcast], *Radio Times* 86: 1114 (2 Feb. 1945), 8.
'The Elusive Classics' (Virgil: *The Eclogues and Georgics*, translated into English Verse by R. C. Trevelyan), *New Statesman and Nation* 29: 724 (5 May 1945), 293.
'"The Hippolytus" of Euripides' [comment on radio adaptation of Gilbert Murray's translation], *Radio Times* 89: 1148 (28 Sept. 1945), 4.
'Mr O'Casey's Memoirs' (*Drums Under the Windows*, Sean O'Casey), *Time & Tide* 26: 45 (10 Nov. 1945), 942.
'L'Écrivain britannique et la guerre', *La France libre* 11: 62 (15 Dec. 1945), 103–9.

'Introduction' to *The Golden Ass of Apuleius*, trans. William Adlington 1566 (London: John Lehmann, 1946), v–ix.

'The Knox New Testament' (*The New Testament*, newly translated into English, by Ronald Knox), *Spectator* 176: 6150 (10 May 1946), 484.

'Pindar: A New Judgment' (*Pindar*, by Gilbert Norwood), *New Statesman and Nation* 31: 795 (18 May 1946), 362.

'*Enter Caesar*' [comment on his play], *Radio Times* 92: 1198 (13 Sept. 1946), 5.

'"Sin and Divine Justice"' [comment on radio adaptation of his translation of Aeschylus' *Agamemnon*], *Radio Times* 93: 1204 (25 Oct. 1946), 13.

'A Greek Satirist who is Still Topical' [comment on his *Enemy of Cant*, radio panorama of Aristophanic comedy], *Radio Times* 93: 1209 (29 Nov. 1946), 5.

'The Traditional Aspect of Modern English Poetry', *La Cultura nel Mondo* (Rome), Dec. 1946, 220–4.

'Scripts Wanted!', *BBC Year Book 1947*, 25–8.

'The Saga of Burnt Njal' [comment on his adaptation of *Njal's Saga* in two parts—*The Death of Gunnar* and *The Burning of Njal*], *Radio Times* 94: 1221 (7 March 1947), 4.

'General Introduction' to his *The Dark Tower and Other Radio Scripts* (May 1947), 9–17.

'Rome: Where History Plays Leap-frog' [comment on his feature *Portrait of Rome*], *Radio Times* 95: 1236 (20 June 1947), 13.

'The English Literary Scene Today: A Return to Responsibility Features the Approach to the Present Crisis', *New York Times Book Review*, 28 Sept. 1947, 1, 34.

'In the Beginning was the Word' [comment on the spoken and the sung word], printed programme note for 'A Recital of Song and Verse by Hedli Anderson, Cyril Cusack, Louis MacNeice': Wigmore Hall, London W1, Sun., 22 Feb. 1948 at 3 p.m., [2].

'An Alphabet of Literary Prejudices', *Windmill* 3: 9 (March 1948), 38–42.

'India at First Sight' [broadcast introduction to a series of features on India and Pakistan, 13 March 1948], *BBC Features*, ed. Laurence Gilliam (1950), 60–4.

'Indian Art' (*Indian Art; The Vertical Man*, by W. G. Archer), *New Statesman and Nation* 35: 888 (13 March 1948), 218.

'Acknowledgement' and 'Note' [on 'The Streets of Laredo'], *Holes in the Sky: Poems 1944–1947* (7 May 1948), 7 and 12.

'*The Two Wicked Sisters*' [comment on his play], *Radio Times*, TV edn, 100: 1292 (16 July 1948), 7.

'Psycho-Moralities and Pseudo-Moralities' [comment on the rebroadcast of his play *The Careerist*, along with its parody *The Life of Subhuman* by Laurence Kitchin], *Radio Times*, TV edn, 100: 1301 (17 Sept. 1948), 7.

'Eliot and the Adolescent', *T. S. Eliot: A Symposium*, comps. M. J. Tambimuttu and Richard March (Sept. 1948), 146–51.

'English Poetry Today' [his broadcast on BBC Eastern Service], *Listener* 40: 1023 (2 Sept. 1948), 346–7.

'*Trimalchio's Feast*' [comment on his adaptation from Petronius' *Satyricon*], *Radio Times* 101: 1314 (17 Dec. 1948), 10.

'Westminster Abbey' [his overseas broadcast], *London Calling*, 23 Dec. 1948, 16–18.

'Experiences with Images', *Orpheus* 2 (1949), 124–32.

'The Crash Landing' [radio feature on India], *Botteghe Oscure* 4 (1949), 378–85.

'Poets Conditioned by their Times' [broadcast of his lecture], *London Calling*, 10 Feb. 1949, 12, 19.

'Portrait of a Modern Man' [comment on his revised play *He Had a Date*], *Radio Times*, TV edn, 102: 1322 (11 Feb. 1949), 7.

'An Irish Proletarian' (*Inishfallen, Fare Thee Well*, by Sean O'Casey), *New Statesman and Nation* 37: 937 (19 Feb. 1949), 184–5.

'Preface' (? Spring 1949) to *Collected Poems 1925–1948* (Sept. 1949), 8, in *Collected Poems*, ed. E. R. Dodds (1966, 1979), xiv.

'Listeners are Warned . . . "a Study in Evil"' [comment on his play *The Queen of Air and Darkness*], *Radio Times* 102: 1328 (25 March 1949), 9.

'An Indian Ride' (*At Freedom's Door*, by Malcolm Lyall Darling), *New Statesman and Nation* 37: 943 (2 April 1949), 334.

'Heart of Byron' (*Don Juan*, by Lord Byron: with an Introduction by Peter Quennell), *Observer*, no. 8235 (3 April 1949), 3.

Letter [correction of Horace quotation in his Byron article in the preceding issue], *Observer*, no. 8236 (10 April 1949), 5.

'Betjeman' (*Selected Poems*, by John Betjeman: chosen with a Preface by John Sparrow), *Poetry* (London), 4: 15 (May 1949), 23–5.

'A Guide-Book for the Educated' (*Fabled Shore*, by Rose Macaulay), *New Statesman and Nation* 37: 953 (11 June 1949), 618.

'The Critic Replies' [comment on Christopher Fry's comedy *The Lady's Not for Burning*], *World Review*, NS, no. 4 (June 1949), 21–2.

'Poetry, the Public and the Critic', *New Statesman and Nation* 38: 970 (8 Oct. 1949), 380–1.

'On Making a Radio Version of *Faust*' [comment on his adaptation of Goethe's *Faust* in six programmes], *Radio Times* 105: 1359 (28 Oct. 1949), 5, 7.

'And so to *Faust*, Part Two', *Radio Times* 105: 1360 (4 Nov. 1949), 6.

'A Poet's Play' (*The Cocktail Party*, by T. S. Eliot), *Observer*, no. 8292 (7 May 1950), 7.

'Great Riches' (*Collected Poems of W. B. Yeats*), *Observer*, no. 8308 (27 Aug. 1950), 7.
'Landscape and Legend' (*The Dreaming Shore*, by Olivia Manning), *Observer*, no. 8311 (17 Sept. 1950), 7.
'Introduction' to *Goethe's Faust, Parts I and II: An Abridged Version*, trans. MacNeice (July 1951), 9–10.
'Athens—City of Contrasts' [comment on his feature *Portrait of Athens*], *Radio Times*, 113: 1462 (16 Nov. 1951), 5.
'The Olympians' (*The Gods of the Greeks*, by C. Kerényi; *Greeks and Trojans*, by Rex Warner), *Observer*, no. 8377 (23 Dec. 1951), 7.
'The Real Mixer' (*Chiaroscuro: Fragments of Autobiography*, by Augustus John), *New Statesman and Nation*, 43: 1100 (5 April 1952), 408.
'The Faust Myth' (*The Fortunes of Faust*, by E. M. Butler), *Observer*, no. 8393 (13 April 1952), 7.
'About Ireland' (*Dublin*, by Maurice Craig; *Ireland and the Irish*, by Charles Duff; *The Emerald Isle*, by Geoffrey Taylor; *The Face of Ulster*, by Denis O'D. Hanna; *Connacht: Galway*, by Richard Hayward), *New Statesman and Nation* 43: 1106 (17 May 1952), 590, 592.
'Sharks Ltd.' (*Harpoon at a Venture*, by Gavin Maxwell), *New Statesman and Nation* 43: 1109 (7 June 1952), 681.
'Notes on the Way', Part 1, *Time & Tide* 33: 26 (28 June 1952), 709–10.
[Acknowledgments], *Ten Burnt Offerings* (11 July 1952), 10.
'Notes on the Way', Part 2, *Time & Tide*, 33: 28 (13 July 1952), 779–80.
'Hatred and Love' (*Rose and Crown*, by Sean O'Casey), *Observer*, no. 8406 (13 July 1952), 7.
'The Bardic Strain' (*Heroic Poetry*, by C. M. Bowra), *New Statesman and Nation* 44: 1121 (30 Aug. 1952), 242.
'Wedding of Simon Karras' [comment on his recordings of a Byzantine wedding ceremony 1950, in the monastery church of Daphni, Greece], *Radio Times* 116: 1504 (5 Sept. 1952), 13.
'Books in General' (*A Reading of George Herbert*, by Rosemond Tuve), *New Statesman and Nation* 44: 1123 (13 Sept. 1952), 293–4.
'A Poet's Choice' (*Poets of the English Language*, edited by W. H. Auden and Norman Holmes Pearson, 5 volumes), *Observer*, no. 8416 (21 Sept. 1952), 8.
'Spenser's Symbolic World' [comment on 12 programmes of *The Faerie Queene*: readings chosen by C. S. Lewis], *Radio Times* 116: 1507 (26 Sept. 1952), 15.
'Yeats's Plays' (*Collected Plays*, by W. B. Yeats), *Observer*, no. 8422 (2 Nov. 1952), 8.
'Portrait of a Would-Be Hero' [comment on his play *One Eye Wild*], *Radio Times*, 117: 1513 (7 Nov. 1952), 6.

Letter, 'Literature and the Lively Arts', *TLS*, 21 Nov. 1952, 761, in reply to an article in *TLS* of 14 Nov. 1952, 749.

Second letter, 'Literature and the Lively Arts', *TLS*, 19 Dec. 1952, 837, in reply to the writer of the article, 5 Dec. 1952, 797.

Untitled article, in *This I Believe*, foreword by Edward Murrow, edited by Edward P. Morgan (1953), 64–5.

'The Sideliner' (*Edward Lear's Indian Journal 1873–75*, edited by Ray Murphy), *New Statesman and Nation* 45: 1152 (4 April 1953), 402.

'Words Are Things Which Ring Out: The Strange, Mighty Impact of Dylan Thomas' Poetry' (*The Collected Poems of Dylan Thomas*), *New York Times Book Review*, 5 April 1953, 1, 17.

'A Greek Story of a Family Curse' [comment on rebroadcast of his adaptation of his translation of Aeschylus' *Agamemnon*], *Radio Times* 119: 1546 (26 June 1953), 21.

'A Poet's Progress' (*W. B. Yeats: Letters to Katharine Tynan*, edited by Roger McHugh), *Observer*, no. 8458 (12 July 1953), 7.

Letter, 'Gordon Her[r]ickx', *New Statesman and Nation* 46: 1168 (25 July 1953), 104.

'Poetry Needs to be Subtle and Tough', *New York Times Book Review*, 9 Aug. 1953, 7, 17.[1]

'A Plea for Sound', *BBC Quarterly*, 8: 3 (Autumn 1953), 129–35.

'The Other Island' (*Mind You, I've said Nothing!*, by Honor Tracy; *The Silent Traveller in Dublin*, by Chiang Yee), *New Statesman and Nation*, 46: 1183 (7 Nov. 1953), 570, 572.

Letter, 'Dylan Thomas Memorial Fund', *The Times*, no. 52788 (25 Nov. 1953), 9d, signed by T. S. Eliot, Peggy Ashcroft, Kenneth Clark, *et al.*, including MacNeice; same letter in *TLS*, 7 Nov. 1953, 762; in *New Statesman and Nation* 46: 1187 (5 Dec. 1953), 719; in *Encounter*, 2: 1 (Jan. 1954); and elsewhere.

'He Weeps by the Side of the Ocean' (*Teapots & Quails*, by Edward Lear), *New Statesman and Nation* 46: 1187 (5 Dec. 1953), 721.

'Dramatising a Tale of a Giant' [comment on his play *The Heartless Giant*], *Radio Times* 122: 1573 (1 Jan. 1954), 4.

'Dylan Thomas: Memories and Appreciations III', *Encounter* 2: 1 (Jan. 1954), 12–13; reprinted in *Dylan Thomas: The Legend and the Poet*, ed. E. W. Tedlock (London: Heinemann, 1960), 85–7, and in *A Casebook on Dylan Thomas*, ed. John Malcolm Brinnin (New York: Crowell, 1960), 282–4.

Note [on *Autumn Sequel*, Canto xviii], *London Magazine* 1: 1 (Feb. 1954), 104.

[1] The original title: 'What Makes a Good Poet'—according to a *New York Times* payment voucher, 11 Aug. 1953, in the possession of Mrs Hedli MacNeice.

Introduction to 'A Dylan Thomas Award', *Dock Leaves* 5: 13 (Dylan Thomas Number, Spring 1954), 6–7 [the prize-winning poem: Anthony Conran, 'For Dylan Thomas (On hearing he was dead)'].

'Round and About Milk Wood' (*Under Milk Wood*, *The Doctor and the Devils*, by Dylan Thomas), *London Magazine* 1: 3 (April 1954), 74–7.

'Greece and the West' (*Fair Greece, Sad Relic*, by Terence Spencer), *New Statesman and Nation* 47: 1210 (15 May 1954), 636.

'In the Grand Manner' (*The English Epic and its Background*, by E. M. W. Tillyard), *New Statesman and Nation* 47: 1215 (19 June 1954), 804.

George Herbert, by Margaret Bottrall; *George Herbert*, by Joseph H. Summers, *London Magazine* 1: 7 (Aug. 1954), 74–6.

The People of the Sea, by David Thomson, *London Magazine* 1: 9 (Oct. 1954), 94, 96.

'Endless Old Things' (*The Letters of W. B. Yeats*, edited by Allan Wade), *New Statesman and Nation* 48: 1230 (2 Oct. 1954), 398.

Prefatory Note to *Autumn Sequel* 1953 (12 Nov. 1954), in *Collected Poems*, ed. E. R. Dodds (1966, 1969), 329.

'I Remember Dylan Thomas', *Ingot* (Steel Co. of Wales), Dec. 1954, 28–30.

'Sometimes the Poet Spoke in Prose: From Squib to Satire—a Collection of the Varied Writing of Dylan Thomas' (*Quite Early One Morning*, by Dylan Thomas), *New York Times Book Review*, 19 Dec. 1954, 1.

Autobiographical sketch, in *Twentieth Century Authors: First Supplement*, ed. Stanley J. Kunitz and Vineta Colby (1955), 624–5.

'Tunnelling into Freedom' [comment on rebroadcast of his play *Prisoner's Progress*], *Radio Times* 126: 1626 (7 Jan. 1955), 5.

'Told in Monologues' [comment on his adaptation of Virginia Woolf's *The Waves* in two programmes], *Radio Times* 126: 1635 (11 March 1955), 5.

Quite Early One Morning, by Dylan Thomas; *The Poetry of Dylan Thomas*, by Elder Olson; *Dylan Thomas*, by Derek Stanford, *London Magazine* 2: 5 (May 1955), 106–9.

'Journey up the Nile' [comment on his feature *The Fullness of the Nile*], *Radio Times* 128: 1651 (1 July 1955), 7.

'The River Nile' [comment on rebroadcast, in series on Africa], *London Calling*, 28 July 1955, 5.

Letter to the Editor, *London Magazine* 2: 9 (Sept. 1955), 69–70, in reply to J. C. Hall's review of *Autumn Sequel*, ibid., 2: 6 (June 1955), 95–7.

Poetry and the Age, by Randall Jarrell; *Inspiration and Poetry*, by C. M. Bowra, *London Magazine* 2: 9 (Sept. 1955), 71–4.

'What Vomit Had John Keats?' (*Dylan Thomas in America*, by John Malcolm Brinnin), *New Statesman and Nation* 51: 1310 (21 April 1956), 423–4.

'The Sound of Bow Bells' [comment on his feature *Bow Bells*], *Radio Times* 131: 1701 (15 June 1956), 5.

'Historic Rouen: a City of Contrasts' [comment on his feature *Spires and Gantries*], *Radio Times* 132: 1707 (27 July 1956), 5.

The Russet Coat, by Christina Keith: *John Clare*, by John and Anne Tibble, *London Magazine* 3: 8 (Aug. 1956), 59–62.

'Ghana: The Birth of an African State' [comment on his feature *The Birth of Ghana*], *Radio Times* 134: 1736 (15 Feb. 1957), 27.

'Indian Approaches' (*Expedition Tortoise*, by Pierre Rambach, Raoul Jahan, and F. Hébert-Stevens; *The Ride to Chandigarh*, by Harold Elvin; *Goa, Rome of the Orient*, by Rémy), *New Statesman and Nation* 53: 1357 (16 March 1957), 346–7.

'Lost Generations?' (*Poetry Now*, edited by G. S. Fraser; *Mavericks*, edited by Howard Sergeant and Dannie Abse), *London Magazine* 4: 4 (April 1957), 52–5.

'Acknowledgements' [with notes on five poems], *Visitations* (10 May 1957), 9.

'Louis MacNeice Writes . . . ' [on his *Visitations*], *Poetry Book Society Bulletin*, no. 14 (May 1957), [1].

'*Nuts in May*' [comment on his play], *Radio Times* 135: 1750 (24 May 1957), 7.

'Fragments I Have Shored' (*Leftover Life to Kill*, by Caitlin Thomas), *New Statesman and Nation*, 53: 1369 (8 June 1957), 741.

'*The Stones of Oxford*' [comment on his feature], *Radio Times* 136: 1767 (20 Sept. 1957), 27.

James Joyce's World, by Patricia Hutchins; *Letters of James Joyce*, edited by Stuart Gilbert, *London Magazine* 4: 11 (Nov. 1957), 73–5.

'Irish Pack Quell the Uppercuts', *Observer*, no. 8690 (19 Jan. 1958), 24.

'A Light Touch under the Stars' [comment on his script for *son et lumière* at Cardiff Castle], *Daily Telegraph*, 9 Aug. 1958, 6.

'Introduction' to *New Poems 1958* [a PEN anthology], ed. Bonamy Dobrée, Louis MacNeice, Philip Larkin (10 Nov. 1958), 9–10.[2]

The Variorum Edition of the Poems of W. B. Yeats, edited by Peter Allt and Russell K. Alspach, *London Magazine* 5: 12 (Dec. 1958), 69, 71, 73, 75.

'Foreword' to *Eighty-Five Poems* (13 Feb. 1959), 7.

'Talking about Rugby', *New Statesman* 57: 1459 (28 Feb. 1959), 286, 288.

'Modern Greeks' (*The Flight of Ikaros*, by Kevin Andrews), *Observer*, no. 8748 (1 March 1959), 19.

'The Battle of Clontarf' [comment on his play *They Met on Good Friday*], *Radio Times* 145: 1882 (4 Dec. 1959), 6.

[2] This introduction was drafted by Larkin and Dobrée only, though signed 'B.D. / P.A.L. / L.MacN.': B. C. Bloomfield, *Philip Larkin: A Bibliography 1933–1976* (1979), 63–4.

'A *Mosaic of Youth*' [comment on his feature], *Radio Times* 145: 1885 (25 Dec. 1959), 37.

'Twin to Drink' (*Report on Rugby*, by W. John Morgan and Geoffrey Nicholson), *New Statesman* 59: 1505 (16 Jan. 1960), 80.

'Being Simple' (*Poetry of This Age: 1908–1958*, by J. M. Cohen), *Spectator* 204: 6868 (12 Feb. 1960), 225–6.

'Eighty Years of Tragi-Comedy' (*Sean O'Casey: The Man and his Work*, by David Krause), *Observer* no. 8804 (27 March 1960), 23.

J. M. Synge, by David H. Greene and Edward M. Stephens; *The Masterpiece and the Man: Yeats as I Knew Him*, by Monk Gibbon, *London Magazine* 7: 8 (Aug. 1960), 70–3.

'*Another Part of the Sea*' [comment on his TV play adapted from his stage play *Traitors in Our Way*], *Radio Times* 148: 1921 (2 Sept. 1960), 2.

'Out of Ugliness' (*Wilfred Owen*, by D. S. R. Welland), *New Statesman* 60: 1545 (22 Oct. 1960), 623–4.

'A Modern Odyssey' (*Poems*, by George Seferis, translated by Rex Warner), *New Statesman*, 60: 1553 (17 Dec. 1960), 978–9.

'When I Was Twenty-One: 1928', *The Saturday Book 21*, ed. John Hadfield (1961), 230–9.

'Louis MacNeice Writes . . . ' [on his *Solstices*], *Poetry Book Society Bulletin*, no. 28 (Feb. 1961), [2].

'That Chair of Poetry', *New Statesman* 61: 1561 (10 Feb. 1961), 210, 212.

'Note', *Solstices* (10 March 1961), 11.

'Growing Up in Ireland' (*An Only Child*, by Frank O'Connor; *Kings, Lords, and Commons*, by Frank O'Connor), *Observer*, no. 8869 (25 June 1961), 25.

'Godot on TV', *New Statesman* 62: 1582 (7 July 1961), 27–8.

'Look at the Faces' [TV review], *New Statesman* 62: 1584 (21 July 1961), 95.

'I Got Those Cathode Blues' [TV review], *New Statesman* 62: 1586 (4 Aug. 1961), 164.

'Pleasure in Reading: Woods to Get Lost In', *The Times*, no. 55162 (17 Aug. 1961), 11b–c.

'Roll on Reality' [TV review], *New Statesman* 62: 1588 (18 Aug. 1961), 225–6.

'In Pursuit of Cuchulain' (*Celtic Heritage*, by Alwyn Rees and Brinley Rees), *Observer*, no. 8878 (27 Aug. 1961), 19.

'Snow and Rock' [TV review], *New Statesman* 62: 1590 (1 Sept. 1961), 284.

'Two Plays, and Spike' [TV review], *New Statesman* 62: 1592 (15 Sept. 1961), 358.

'Pins and Needles' [TV review], *New Statesman* 62: 1594 (29 Sept. 1961), 450–1.

'One-Eyed Hero' [comment on his revised play *One Eye Wild*], *Radio Times*, 153: 1983 (9 Nov. 1961), 30.

'Come in, Dominic' (*Teems of Times and Happy Returns*, by Dominic Behan), *New Statesman* 62: 1602 (24 Nov. 1961), 795–6.

'A Fleet Street Chronicle' [comment on his play *Let's Go Yellow*], *Radio Times* 153: 1988 (14 Dec. 1961), 26.

Letter '"Let's Go Yellow"', *Listener* 67: 1710 (4 Jan. 1962), 33.

'Quiet Irish Charm' (*Twice Round the Black Church*, by Austin Clarke), *Sunday Telegraph*, 4 Feb. 1962, 7.

'Nine New Caps' [rugby], *New Statesman* 63: 1614 (16 Feb. 1962), 239–40.

'Goethe's *Faust*' [comment on his adaptation abridged], *Radio Times* 154: 2000 (8 March 1962), 50.

'*The Mad Islands*' [comment on his play], *Radio Times* 154: 2003 (29 March 1962), 39.

'C'est la terre' (*The Furrow Behind Me*, by Angus MacLellan; *The Hard Road to Klondike*, by Michael MacGowan; *My Ireland*, by Kate O'Brien), *New Statesman* 63: 1632 (22 June 1962), 911–12.

'Under the Sugar Loaf' [Dublin], *New Statesman* 63: 1633 (29 June 1962), 948–9.

'The Two Faces of Ireland' (*Brendan Behan's Island*, by Brendan Behan; *West Briton*, by Brian Inglis), *Observer*, no. 8935 (30 Sept. 1962), 29.

'Blood and Fate' (*The Odyssey*, translated by Robert Fitzgerald; *Patrocleia*, Book XVI of Homer's *Iliad*, adapted by Christopher Logue), *Listener* 68: 1749 (4 Oct. 1962), 527.

A Preface to 'The Faerie Queene', by Graham Hough, *Listener* 69: 1766 (31 Jan. 1963), 213.

'Yeats at Work' (*Between the Lines*, by Jon Stallworthy), *Listener* 69: 1773 (21 March 1963), 521.

'The Ould Opinioneer' (*Under a Colored Cap*, by Sean O'Casey), *New Statesman* 65: 1677 (3 May 1963), 678–9.

'Gael Force at Wembley' [rugby and Gaelic football], *New Statesman* 65: 1682 (7 June 1963), 876.

'William Empson', *The Concise Encyclopedia of English and American Poets and Poetry*, ed. Stephen Spender and Donald Hall (1963), 127–8.

The Saga of Gisli, translated by George Johnston, *Listener* 69: 1787 (27 June 1963), 1083.

'Great Summer Sale' [report on President Kennedy's visit to Ireland], *New Statesman* 66: 1686 (5 July 1963), 10, 12.

'Frost' (*The Poetry of Robert Frost*, by Reuben Brower), *New Statesman* 66: 1687 (12 July 1963), 46.

'*Persons from Porlock*' [comment on his play], *Radio Times* 160: 2076 (22 August 1963), 44.

'Louis MacNeice Writes . . .' [on his *The Burning Perch*], *Poetry Book Society Bulletin*, no. 38 (Sept. 1963), [1].

'Introduction' to his *Christopher Columbus*, Faber school edition (25 Sept. 1963), 7–10.

'Childhood Memories', from an unscripted talk [recorded in Belfast, July 1963] on the Third Programme, *Listener* 70: 1811 (12 Dec. 1963), 990.

'Introduction' to his *The Mad Islands* and *The Administrator* (13 March 1964), 7–9.

'Notes' to his *One for the Grave* (29 Jan. 1968), 13–14.

Addenda[1]

Lysistrata, Aristophanes: *A new English version by Reginald Beckwith and Andrew Cruickshank* (Gate), *Time & Tide*, 17: 12 (21 March 1936), 422.

Love's Labour's Lost (Old Vic), *Time & Tide* 17: 38 (19 Sept. 1936), 1294.

Charles the King, Charles Colbourne (Lyric), *Time & Tide* 17: 42 (17 Oct. 1936), 1458.

Plot Twenty-one, Ronald Ackland (Embassy), *Time & Tide* 17: 43 (24 Oct. 1936), 1494.

Muted Strings, Arthur Watkyn (Daly's), *Time & Tide* 17: 46 (14 Nov. 1936), 1616.

The Housemaster, Ian Hay (Apollo), *Time & Tide* 17: 47 (21 Nov. 1936), 1646.

Your Number's Up, Diana Morgan, Robert MacDermot, and Geoffrey Wright (Gate), *Time & Tide* 18: 1 (2 Jan. 1937), 24.

The Tobacco Road, Jack Kirkland: *Based on the novel by Erskine Caldwell* (Gate), *Time & Tide* 18: 22 (29 May 1937), 738.

Bonnet Over the Windmill, Dodie Smith (New), *Time & Tide* 18: 38 (18 Sept. 1937), 1247.

Of Mice and Men, John Steinbeck (Gate Theatre), *Time & Tide* 20: 16 (22 April 1939), 528. [Produced by Norman Marshall.]

[1] These notices and reviews (1936–9) are signed simply 'L.M.', apparently by Louis MacNeice. Later notices (1958–60) signed 'L.M.' are by Robert Eaton: see 'Book Contents', *Time & Tide* 39: 49 (6 Dec. 1958), 1498.

Index